Drinking Water Quality Management from Catchment to Consumer

Drinking Water Quality Management from Catchment to Consumer

A practical guide for utilities based on Water Safety Plans

Editor
Bob Breach

IWA
Publishing
London • New York

Published by **IWA Publishing**
Alliance House
12 Caxton Street
London SW1H 0QS, UK
Telephone: +44 (0)20 7654 5500
Fax: +44 (0)20 7654 5555
Email: publications@iwap.co.uk
Web: www.iwapublishing.com

First published 2012

British Library Cataloguing in Publication Data
A CIP catalogue record for this book is available from the British Library

Library of Congress Cataloging-in-Publication Data
A catalog record for this book is available from the Library of Congress

ISBN13: 9781843393863
ISBN: 1843393867

Contents

Appendix 1

Appendix 2

Foreword

Steve E Hrudey

Dr. John Snow's rational scientific hypothesis about how cholera was spread by sewage-contaminated drinking water in urban London of the mid 1800s provided one of the first critical insights that ultimately led to enormous public health benefits. A key aspect of these advances was developing an understanding of the true cause of waterborne disease to allow development of effective water treatment and distribution, mainly by filtration and disinfection. Nearly half of the total mortality reduction, three-quarters of the infant mortality reduction and nearly two-thirds of the child mortality reduction observed in major U.S. cities in the early 20th century has been attributed to clean water (Cutler and Miller 2004). These achievements are foundational examples of how understanding and knowing the key elements of a drinking water system have yielded enormous progress.

Most residents of affluent countries, now with remarkable reliability, are delivered fresh, clean, safe drinking water on demand for a tiny fraction of their disposable income. When we contrast this reality with our past, or the current reality throughout the impoverished majority of the planet, our drinking water supplies in the developed world are surely one of the most valuable advances of our age. At the same time, this achievement is likely to be one of the most under-valued and least appreciated conveniences which too many of us simply take for granted. Unfortunately, such complacency has been a common root cause for many of the waterborne outbreaks which continue to occur in nations whose affluence is difficult to reconcile with the causes of such outbreaks (Hrudey and Hrudey 2004). We need to avoid becoming victims of our own success by allowing complacency over our demonstrable achievements to undermine the level of drinking water safety that some have come to take for granted.

Looking back on my career, I can recall in 1970 during my water treatment coursework for an MSc in Public Health Engineering that our water treatment instructor told us there was no need for further research in drinking water because we already knew everything we needed to know for assuring safe drinking water. Of course, since then we have had numerous important discoveries including revelations in 1971 about *Giardia lamblia* being a cause of drinking water outbreaks in water treatment plants with full conventional water treatment, in 1974 about trihalomethanes being the first disinfection by-products to be documented, in 1976 when *Cryptosporidium parvum* became recognized as a human pathogen and as a cause of waterborne outbreaks in 1984, in 1977 when *Campylobacter* spp., now recognized as one of the most common causes of bacterial drinking water outbreaks, was first determined to be a human pathogen, and in 1977 when arsenic was found to be associated with *"black foot disease"* and ultimately skin cancer in Taiwan drinking water and more recently in Bangladesh. Of course, we now face new revelations every year of possible threats to human health from trace substances in drinking water, so much so that drinking water providers need considerable experience and judgment to successfully keep their focus on what potential contamination source truly poses a serious threat to their drinking water safety.

Providing safe drinking water is a complex, challenging business, yet many of the basic elements are deceptively simple. The combination of apparent simplicity overlying serious, inevitable challenges makes for a situation which gives rise to failure unless rigorous, thorough and consistent commitment to safety is fostered and promoted.

While our advances in assuring safe drinking water are impressive, failure remains only a few mistakes away. Microbial pathogens, the most prevalent risk to drinking water safety, are found wherever humans, their pets and livestock and/or wildlife are found. In other words, microbial pathogens are a pervasive risk to all drinking water systems if they are allowed to contaminate water at any point from the source water to the consumer's tap. Our modern populated society contributes multiple impacts with pervasive and numerous contaminant risks to source water. We can do a perfect job of dealing with these only to encounter treatment process upsets or contamination of treated water storage or the distribution system.

Some water professionals may be too content with what we have achieved and not enough concerned with the number of things which can go wrong, sometimes in an almost unimaginable manner. For example, recent experience in an advanced U.K. water treatment plant providing pre-ozonation, clarification, filtration, ozonation, granular activated carbon (GAC) adsorption and chlorination allowed a small outbreak of cryptosporidiosis among consumers. This failure was detected, but not prevented by continuous monitoring for oocysts in treated water (Sinclair and Lightbody 2008). Ultimately, the cause was found to be an adventurous rabbit which had worked its way into a GAC backwash tank where it most likely drowned during backwash before being washed into the chlorination tank where it contaminated the treated water with a strain of oocysts not previously known to infect humans. This failure in a system which had gone beyond conventional treatment, monitoring and response practices highlights how challenging are the measures necessary to assure safe drinking water.

Because drinking water safety effectively demands assuring negligible health risk to consumers, best practice has evolved for implementing and maintaining an effective multiple barrier approach to minimize the chance of contaminants reaching consumers in a concentration capable of causing adverse health outcomes. The water safety plan approach advocated in this book is simply the means for assuring that the multiple barrier approach is functioning effectively.

Of course, drinking water quality criteria, as captured in documents such as the WHO Drinking Water Guidelines, continue to be essential for judging whether levels of contamination pose a health risk. As such, they are essential elements of assuring drinking water safety. However, a critical insight about drinking water quality criteria among water safety plan proponents is that focusing primarily on monitoring of treated water for compliance with water quality criteria is inherently reactive because such compliance monitoring provides too little information, too late.

Water safety planning can shine a critical light on the greatest vulnerability in a drinking water system. The insights gained from water safety planning can provide a water authority with the additional knowledge, capacity and resilience to face the many challenges to assuring safe drinking water in its own system.

This book, written by several knowledgeable and experienced authors, provides an excellent introduction and elaboration of the rationale and means for adopting and implementing water safety plans by modern water authorities. As such, this book should be an essential addition to the management library of all forward-looking drinking water providers.

REFERENCES

Cutler, D. M., and G. Miller 2004. "The Role of Public Health Improvements in Health Advances: The 20th Century United States." National Bureau of Economic Research. NBER Working Paper No. 10511. http://www.nber.org/papers/w10511

Hrudey, S. E., and E. J. Hrudey. 2004. Safe Drinking Water – Lessons from Recent Outbreaks in Affluent Nations. IWA Publishing, London. 514pp.

Sinclair, M. and P. Lightbody. 2008. DWI Outbreak Investigation. *Health Stream*. #52 - December 2008. http://www.wqra.com.au/_dyn/media/r376/system/attrib/file/317/HS52.pdf

About the authors

Bob Breach MBE, BSc, PhD, MCIWEM: Former Head of Quality and Environment for a large UK water utility, former chairman Eureau water supply commission, co-author of the Bonn Charter. Now an independent consultant. (bob.breach@btinternet.com).

Ricardo Hirata DSc, MSc, PGeo: Hydrologist, World Bank Groundwater Management Advisory Team (GWMATE), Professor and Director of Groundwater Research Centre (CEPAS), University of São Paulo, Brazil. (rhirata@usp.br).

Steve E. Hrudey FRSC, FSRA, IWAF, PhD, DSc (Eng), PEng: Professor Emeritus, Analytical & Environmental Toxicology, Faculty of Medicine & Dentistry, University of Alberta, Canada. (steve.hrudey@ualberta.ca).

Yahya Omar MSc: Research student within Centre for Water Science at Cranfield University, UK. (y.y.omar@cranfield.ac.uk).

Simon Pollard DSc: Professor and Dean of the Faculty of Environment, Science and Manufacturing, Cranfield University, UK. (s.pollard@cranfield.ac.uk).

Chris Rockey BSc: Science and Water Quality Manager, South West Water, UK. (crockey@southwestwater.co.uk).

Rui Sancho MSc, MBA: Coordinator of the Department of Drinking Water Operations in the Western Algarve and Coordinator of the Water Safety in Águas do Algarve, Portugal. (r.sancho@aguasdoalgarve.pt)

Michael Storey BSc, PhD: Customer Research and Strategy Manager, Sydney Water, Australia. (michael.storey@sydneywater.com.au).

Corinna Summerill BSc, MSc, PhD: Former research student within Centre for Water Science at Cranfield University, UK. (corinnasummer@yahoo.co.uk).

José Manuel Pereira Vieira BSc, MSc, PhD: Former Vice-Rector of University of Minho, Vice-President of the Portuguese Association of Engineers, Professor and Head of the Civil Engineering Department, University of Minho, Portugal. (jvieira@civil.uminho.pt).

Tom Williams BSc, MSc: Programmes Manager, International Water Association, The Netherlands. (Tom.Williams@iwahq.org).

Acknowledgements

This book would not have been possible without the hard work and expert input from many people.

Firstly, Tom Williams of IWA who not only sponsored the book and wrote a chapter, but also led the Bonn Network and continues to help promote the effective management of drinking water quality around the world. Also Steve Hrudey for graciously accepting an invitation to write the foreword, and for his seminal book which reminds us all about the need for constant vigilance in managing drinking water quality.

Secondly, to my co-authors who, in addition to their busy professional commitments, have worked against difficult deadlines to write their chapters. My deep thanks go to Ricardo Hirata, Yahya Omar, Simon Pollard, Chris Rockey, Rui Sancho, Corinna Summerill, Michael Storey and Jose Vieira.

Thirdly, to the members of the Bonn Network with whom I have worked closely over the last few years and who have provided expert help and/or peer reviews of the technical content of the book. Their practical insight and support has been invaluable. Particular thanks go to: Maria João Benoliel, Joana Pinto Coehlo, Ines Freitas, Marta Ganzer, John Howard, Christopher Kanyesigye, Jean-Francois Loret, Carla Morais, Kista Naidoo, Yvonne Nijdam, James Pratt, Andre Luis Gois Rodrigues, Pauline Adams-Russell, Jaffaran Suhaimi, Sarah Tibatemwa, Chris Viljoen, Richard Walker and Azuan Zahdi. I have also had great support from a number of expert colleagues who have peer reviewed specific chapters. Specifically I am most grateful to: Urmila Brighu, Joby Boxall, Don Bursill, Nick Everall, Martin Furness, Tom Hall, Bob Harris, Rob Keirle, Mike Porter, and Mick Whelan. Technical help and/or figures have also been provided by Lisa Barrott, Tim Darlow, Rick Gelting, Asoka Jayaratne and Sue Jenkins,

Grateful thanks are also due to Maggie Smith and the team at IWA Publishing for the professional support and encouragement I have received.

In writing and editing this book I have used knowledge and inspiration obtained from a great many colleagues in many different countries with whom I have worked over many years. The countless hours of discussion and stimulation have been a privilege for which I will be eternally grateful.

Finally my love and thanks to my wife Janet, who has not only helped me with diligent proof reading of this book, but has provided unfailing support and encouragement throughout my entire career and many long periods away from home.

Bob Breach

Background and use of this book

The primary goal of the IWA Bonn Charter is the provision of:

"*Good safe drinking water which has the trust of consumers*"

To secure this goal the Charter provides a set of high level principles which are universally applicable to both developed and developing countries across the world. At the heart of the process is the development of a risk based water safety plan from catchment to consumer. But the document also covers a number of other important areas including the setting of water quality standards, roles and responsibilities, institutional arrangements, communication, and water pricing and financing. Central to the idea of the Bonn Charter is that whilst the principles of good water quality management are universal, the way they are applied locally will depend on a range of factors including cultural, legal, institutional, and socioeconomic.

Since the joint launch in 2004 of the 3rd edition of the WHO Guidelines for drinking water quality and the IWA Bonn Charter, both WHO and IWA have worked in close partnership to support continuous improvement in the management of drinking water quality. As part of its contribution to this work IWA created the Bonn Network – a global network of water suppliers committed to providing good safe drinking water. This best practice book has evolved out of the work of the Bonn Network. The emphasis is on practical information "*by operators for operators*", learning from the experience of a range of water suppliers and other experts across the world. It aims to provide practical guidance to help utilities improve their management of drinking water quality. It focuses on piped water distribution systems, not small local community or household systems which are well covered in excellent publications by WHO and others.

The book is based on the Water Safety Plan approach and the principles of the Bonn Charter for Safe Drinking Water, and is intended to complement and support other relevant publications, particularly the WHO Water Safety Plan Manual. It encourages users to think through the best way for them to improve drinking water quality management, taking account of local factors. It recognises that whilst water suppliers are central to the provision of good drinking water, this has to take place within a broader legal, institutional, and financial framework which is the responsibility of Government and many others. It thus highlights the ways that water suppliers can work with a range of partners to ensure that all are committed to the same goal of good safe drinking water.

Throughout the book the term "*consumer*" has been used rather than "*customer*", since this covers a much wider range of individuals. This includes anybody that uses, or comes into contact with, drinking water either directly or indirectly, and can include home owners, residents, employees, operators, or any other water user in a particular building or site.

The book is structured in a format which allows readers to easily identify a particular topic of interest and is facilitated by a comprehensive contents section. It is hoped that this approach will be helpful to all water suppliers, whether they are new to water safety plans, or already have significant experience in their implementation. It should, therefore, allow readers to use the various chapters to identify gaps in their procedures for managing drinking water quality, so that they can be progressively improved on a prioritised basis which meets their needs.

Additional information can be obtained through the provided references at the end of each Chapter, and two important web based information resources:

- The World Health Organisation site on water safety plans (http://www.who.int/water_sanitation_health/dwq/WSP/en/index.html)
- The WHO/IWA water safety plan portal (http://www.wsportal.org/ibis/water-safety-portal/eng/home)

The main issues covered in each Chapter are as follows:

- ***Chapter 1*** "Why water safety plans and the Bonn Charter" describes the reason for the development of WSPs and the Bonn charter. It highlights the benefits of the WSP approach and how this links to the setting and achievement of drinking water quality targets, including not only the safety of drinking water, but its acceptability to consumers.
- ***Chapter 2*** "Developing a supportive organisational culture" describes how improving the risk management culture of an organisation can support the development of water safety plans and suggests practical approaches to help utilities develop such an approach.
- ***Chapter 3*** "Operational policies to support water safety plans" summarises the principles of risk management and then describes some of the key operational policies and practices which are necessary to support effective development of water safety plans. This includes asset management, training and awareness, operational procedures, emergency planning, training, purchasing and contractor management, and monitoring, documentation and reporting.
- ***Chapter 4*** "Developing a catchment water safety plan" sets out some of the factors underpinning the development of a catchment WSP, provides advice on developing a catchment water safety plan in conjunction with external stakeholders, describes methodologies for assessing and managing both surface and groundwater catchments and briefly summarises common catchment risks and how they might be mitigated.
- ***Chapter 5*** "Developing a treatment water safety plan" describes some of the issues that need to be considered when developing a WSP for water treatment plan. It highlights the importance of an overall multi barrier approach and then reviews typical water quality risks, mitigation measures and control points for a range of typical water treatment technologies. Finally, the importance of site security, and plant monitoring and control are described along with approaches to setting operational water quality targets to ensure that the final water quality delivered to consumers is acceptable.

- ***Chapter 6*** "Developing a distribution water safety plan" overviews each generic area, activity or event within the distribution network that could pose a risk to water quality, and summarises typical mitigation measures that might be adopted and suitable control points.
- ***Chapter 7*** "Developing a consumer water safety plan" provides general guidance to water utilities on how to work with consumers and relevant stakeholders to promote the adoption of consumer WSPs in both residential and commercial settings. This includes how to, identify relevant stakeholders and their roles and responsibilities, assess and mitigate common risks that are associated with the consumer's water system, identify consumer groups particularly at risk, provide effective consumer communication tools and assess water acceptability to consumers.
- ***Chapter 8*** "Global experience of WSP implementation" provides insight into some of the experiences of implementing WSPs over the past few years and aims to distil and synthesise some of the common challenges, critical success factors and opportunities for WSP implementation. This includes lessons learned from implementation, monitoring and evaluation, understanding the inter-linkages between WSPs and other utility planning and management processes, and impact at scale from global implementation of WSPs.

The underlying basis of a water safety plan is to assess all risks to water quality from catchment to consumer in a structured way; put in place appropriate controls to mitigate these risks; and verify the effectiveness of the controls. Chapters 4-7 describe the very large number of potential hazards and hazardous events from catchment to consumer that *could* pose a risk to the quality of water received by consumers, and there are likely to be other hazards not described in the book which may also need to be assessed. However, the relevance and actual level of risk for any particular water utility can only be assessed locally. It is also important to remember that hazards and risks can change over time so need to be continually reviewed.

One of the difficulties commonly experienced by water utilities when developing a water safety plan for the first time is dealing with the large number of potential risks that are initially identified. This emphasises the importance of a rigorous risk prioritisation process. As set out in Chapter 3, whilst a water supplier will wish to take all practicable steps to minimise risks to water quality, particularly those impacting public health, it is unrealistic to assume that all risks, however low, can be completely avoided. Indeed, many lower level risks will not be able to be addressed, at least in the short to medium term. Thus, once a list of all potential risks has been identified, a process will have to be put in place to rank these risks for action according to priorities identified by utility management.

The number of potential hazards and hazardous events that could pose a risk to water quality does, though, highlight the scale of the challenge in producing consistently good quality drinking water and the need for continued vigilance by all water utility staff and the many stakeholders with which they have to work. But this should not be seen as a barrier to implementation. WSPs are scalable, and it is better to do what you can with what you have, which can still yield big improvements, rather than thinking "*we don't have the resources so we won't do it*". This approach has been summarised by WHO/IWA as "*Think big, start small, scale up*". It is hoped that this book will be of help to utilities in managing the vital task of delivering safe drinking water to their consumers.

Bob Breach

Glossary of key terms and abbreviations

This glossary provides an explanation of frequently used terms and abbreviations used in this book. Wherever possible they are derived from those used in the WHO Water Safety Plan Manual or other WHO publications.

AOC: Assimilable organic carbon.

AwwaRF: American waterworks association Research Foundation now known as Water Research Foundation.

ADWG: Australian Drinking Water Guidelines.

AWWA: American Water Works Association.

Bonn Charter for Safe Drinking Water: The IWA Bonn Charter for Safe Drinking Water provides a framework outlining the institutional, managerial and operational requirements for the effective provision of safe drinking water (see Appendix 1).

Catchment: In this book the term catchment is used to describe the whole area of land from which water drains to a point, either on the surface or underground, from which it is abstracted for use in the production of drinking water. It covers surface waters such as rivers/canals and impounded/pumped reservoirs, groundwater, and coastal/brackish areas.

Control: To take all necessary measures to ensure and maintain compliance with criteria established in a WSP. The state wherein correct procedures are being followed and criteria are being met.

Control measure: Any action and activity that can be used to prevent or eliminate a water safety hazard or reduce it to an acceptable level.

Control Point: A step at which control can be applied to prevent or eliminate a water safety hazard or reduce it to an acceptable level. Some might be termed critical (key) control points (CCP) at which control might be essential to prevent or eliminate a particularly serious water safety hazard.

Corrective Action: Any action to be taken when the results of monitoring at the control point indicate a loss of control.

Critical Limit: A criterion which separates acceptability from unacceptability.

DBPs: Disinfection byproducts

Distribution System: In this book the term distribution system covers any piped network used to deliver potable water to consumers.

DOC: Dissolved organic carbon

EUREAU: European Federation of National Associations of Water and Wastewater Services

Flow diagram: A systematic representation of the sequence of steps or operations from catchment to consumer used in the production of drinking water.

GAC: Granular activated carbon.

Grey water: Water from a kitchen, bath, or laundry which normally does not contain significant concentrations of excreta.

HACCP: Hazard Analysis and Critical Control Point.

Hazard: A biological, chemical, physical or radiological element that has the potential to cause harm. Another word for hazard is "contaminant".

Hazardous event: An incident or situation that can lead to the presence of a hazard.

Hazard Analysis: The process of collecting and evaluating information on hazards and conditions leading to their presence to decide which are significant for water safety and therefore should be addressed in a water safety plan.

HPC: Heterotrophic plate count.

NHMRC: National Health and Medical Research Council (Australia).

NOM: Natural organic matter.

Operational monitoring: The act of conducting a planned series of observations or measurements of control parameters to assess whether a control measure is operating within its design criteria.

Reliability: The term reliability refers to both the quality of water supplied, and the dependability of supplies.

Risk: The likelihood of identified hazards or hazardous events causing harm to the exposed population in a specified timeframe and the magnitude of that harm and/or its consequences.

Risk assessment: For the purposes of this book, risk assessment has the same meaning as hazard analysis.

Risk Score: The score assigned to a hazard based on the risk analysis process.

SCADA: Supervisory control and data acquisition.

SOP: Standard operating procedure.

THM: Trihalomethanes.

TOC: Total organic carbon.

USEPA: United States Environmental Protection Agency.

Validation: Obtaining evidence that the WSP is working effectively and that the performance of control measures used for the WSP can effectively meet the water quality targets set.

Verification: The application of analysis methods, procedures, and other tests, in addition to operational monitoring, to determine whether the water quality objectives of the WSP are being met. That is checking whether the system is delivering water of the desired quality and whether the WSP is being implemented in practice or needs modification.

Water Resources: In this book, the term water resources refers to raw water collection, whether from surface sources, groundwater, the sea or other brackish sources.

Water Suppliers: The term Water Suppliers should be taken to include all public and private suppliers of potable water provided through a piped network.

WHO: World Health Organisation.

WSAA: Water Services Association of Australia.

WSP: Water safety plan.

Dedication

This book is dedicated to all those around the world who work ceaselessly, and often under difficult conditions, to consistently produce and distribute good safe drinking water to their consumers. This is a vitally important, but often undervalued task, although one which is essential for the health and well being of communities.

Chapter 1

Why water safety plans and the Bonn Charter?

Bob Breach

1.1 INTRODUCTION

The launch in 2004 of the 3rd edition of the WHO guidelines for drinking water quality (WHO 2004) jointly with the IWA Bonn Charter for Safe Drinking Water (IWA 2004) marked the culmination of a huge shift in worldwide advice on water quality management that had been evolving for some years. This redressed the balance of effort away from reliance solely on end product testing towards a more proactive risk based approach.

The WHO Guidelines stated that "*The most effective means of consistently ensuring the safety of a drinking-water supply is through the use of a comprehensive risk assessment and risk management approach that encompasses all steps in water supply from catchment to consumer. In these Guidelines, such approaches are called water safety plans (WSPs)*".

This was driven by a growing recognition that safe drinking water could only be consistently secured through such a risk based approach, thus contributing to the Millennium Development goal to "*Halve, by 2015, the proportion of the population without sustainable access to safe drinking water and basic sanitation*". Despite good progress with improving drinking water across the world much remains to be done (United Nations 2010, Pruss-Ustun and Corvalan 2006). Furthermore, even in developed countries serious outbreaks of waterborne disease continue to occur which highlight the need for ongoing vigilance in the management of water supply systems (Hrudey and Hrudey 2004).

Since 2004 WHO and IWA have been jointly promoting the WHO drinking water guidelines in parallel with the Bonn Charter and have developed a number of strategies to secure high levels of their adoption, including providing a wide range of valuable guidance material. This has been strongly supported by great efforts from governments, water suppliers and other stakeholders across the world to put into practice such a risk based approach for water quality management. The importance of this was reinforced by the adoption by the World Health Assembly in May 2011 of a major Resolution on Drinking Water, Sanitation and Health (WHO 2011). The WHO drinking guidelines have also been subject to a process of continuous updating and the latest 4th edition was published in July 2011 (WHO 2011b).

1.2 WHY AND HOW DID THE WATER SAFETY PLAN APPROACH DEVELOP?

Understanding of the importance of risk based approaches to the management of drinking water quality is actually not new. Over 100 years ago a paper (Horton 1898) in the UK journal "*Proceedings of the Institute of Sanitary Engineers*" said "*When the supply is from shallow wells the water is generally uncertain and therefore unsafe to drink; but deep wells are by no means beyond suspicion.* It illustrated this by description of a source near to London *"The water was from a deep well... but when it was opened for inspection the source of impurity was very evident ... about 20 ft (6.5 m) of the upper part was built of loose brickwork and every facility was thus afforded for draining the manured ground into the well*".

In 1970 the WHO publication "*European Standards for Drinking Water*" (WHO 1970) identified that *"No bacteriological or chemical examination however careful can take the place of a complete knowledge of the conditions at the source of supply and throughout the distribution system"*. Importantly, it also reminded us that "*Every supply should be regularly inspected from source to outlet by experts, and sampling, particularly for bacteriological examination, should be repeated under varying climatic conditions, especially after heavy rainfall and after major repair or construction work"*.

A few years later WHO published a monograph (WHO 1976) on the "*Surveillance of Drinking Water Quality*" which identified a range of activities necessary to ensure safe drinking water for water suppliers with different levels of resources. This included sanitary surveys, training of staff, reporting requirements, remedial action following failure, design standards/criteria, control of cross connection, plumbing codes, and materials and additive standards. Most interestingly, in Annexe 6 of that publication, there was a model water supply sanitary survey form that could now be seen as a very early example of a water safety plan. For example, as well as water quality monitoring it identified the need for assessment, documentation and control of a variety of key risks from catchment to consumer.

Thus, even many years ago there was a recognition that regular risk assessment of the whole water supply system from catchment to consumer, and implementation of suitable control measures, was a vital part of keeping drinking water safe. However, in the 1980s and 1990s, at least in the developed world, water quality management and the setting of water quality standards was increasingly being driven by an explosion in the capability of water analytical technology. This led to an increasingly wide ranging number of different substances being detected in water, albeit mostly at relatively low levels. This then opened a debate about how and to what level these substances should be controlled. And this in turn tended to lose focus on some of the basic risk prevention principles that had stood water utilities in good stead for many years. In many utilities a culture developed which based success solely on the outcome of laboratory testing and this also meant that the need for a proactive risk assessment mitigation approach was lost or downplayed.

In the mid 1990s the idea of more structured risk based management systems for drinking water quality began to emerge. One of the earliest references to the approach was a paper (Havelaar 1994) entitled "*Application of HACCP to drinking water supply*". HACCP (Hazard Analysis and Critical Control Point) had been used for managing risks to product quality extensively in other sectors, particularly the food industry (Van Schothorst 2004). However, the approach had hitherto not been seriously considered for the water sector.

In parallel a discussion had been taking place within the Australian water industry about a new risk based approach to managing drinking water quality. In 1996 the National Health and Medical Research Council published the Australian Drinking Water Guidelines (NHMRC 1996). In addition to updating the Australian guidance and standards related to drinking water quality, the document endorsed for the first time the use of quality management systems; a forerunner of what was later to be called Water Safety Plans. Subsequently this led to the adoption of a framework for managing drinking water quality which was incorporated into the Australian Drinking Water Guidelines in 2004 (NHMRC 2004). See Figure 1.1 below.

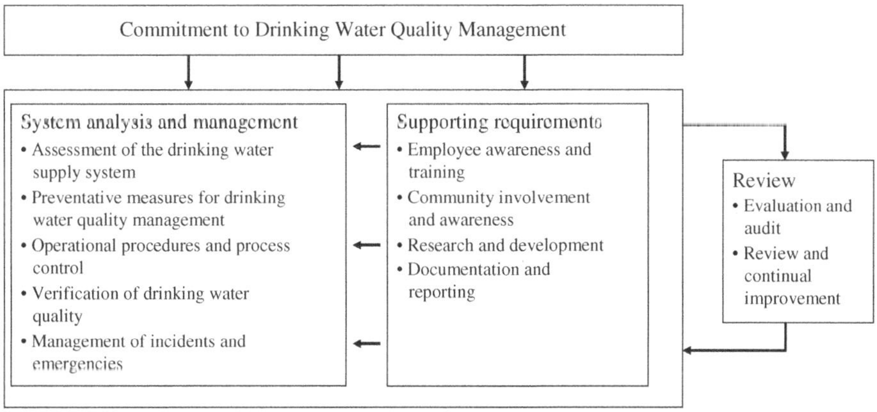

Figure 1.1 Australian framework for management of drinking water quality 2004

At the other side of the world Icelandic waterworks first began using HACCP as a preventative approach for water safety management in 1997. Since then implementation has been ongoing and currently about 68% of the Icelandic population receive drinking water from waterworks with a water safety plan based on HACCP (Gunnarsdottir and Gissurarson 2008).

The potential to increase consistency in approaches to assessment and management of water-related microbial hazards was discussed by an international group of experts coordinated by WHO between 1999 and 2001. These discussions led to the development of a harmonised framework (The Stockholm Framework) which was intended to inform the process of development of guidelines and standards. Subsequently this led to the publication "*Water quality - Guidelines, standards and health - Assessment of risk and risk management for water related infections and diseases*" (Fewtrell and Bartram 2001). Among a range of topics this included a chapter on risk management strategies including examples of the use of HACCP principles adapted to drinking water quality. In parallel with this initiative, WHO had begun the massive task of updating its highly influential 2nd edition of "*Guidelines for Drinking Water Quality*" which had last been issued in the mid 1990s. At that time the need for a process of rolling review had been identified and in 2000 a plan was set in place to develop the 3rd edition of the guidelines.

In February 2001 drinking water supplier associations from Australia (WSAA), the USA(AwwaRF) and Europe (EUREAU) invited a group of experts to a workshop to discuss "*A framework for assuring the quality of drinking water in the 21st century*" which was held in Bonn, Germany. The workshop was hosted by the German Water Association and, importantly, supported by regulatory agencies as well as WHO and other scientific and public health experts. The workshop agreed that the primary goal of the water supply function was:

"To provide good safe drinking water which has the trust of consumers"

To meet this goal a number of key principles were identified.

- *Good safe drinking water can only be provided through an integrated approach from catchment to consumers' taps.*

- *Such an integrated approach will require close cooperation and partnership between Governments, water suppliers, health agencies, environmental agencies, land users, contractors, plumbers and consumers themselves.*
- *Transparency of the quality assurance process, including the derivation of standards, is vital for consumer confidence.*
- *It is believed that a common framework for assuring drinking water quality can be developed, but this must be:-*
  - *Based on best available scientific and medical advice.*
  - *Place greater emphasis on proactive and prevention based quality management systems.*
  - *Sufficiently flexible to take account of different legal, institutional, cultural, and socio-economic situations in different countries.*

In describing the new approach to quality management systems the report helpfully clarified the distinction between what were then termed "Output control" (i.e. water quality monitoring) and "Input control" (i.e. management procedures). It explained that "*traditionally the verification of processes to ensure drinking water quality has been by measuring the value or concentration of various parameters and comparing the results to predetermined standards or guidelines.*" It added that "*Such output control values will continue to play a critical role in assuring drinking water quality, but in future there should be a much greater emphasis on the complementary use of Input Control Systems.*" Input control systems were identified as having three key elements that underpin effective quality assurance:

- *A system wide, risk based assessment of safety from catchment to consumer's tap.*
- *Identification of the most effective control points to reduce risk.*
- *Effective operational, maintenance and management plans to deal with both routine and abnormal operating conditions.*

This must be supplemented by:

- *Measures to assess the effectiveness of the controls that have been put in place.*
- *Independent third party verification of the effectiveness of the control system although this should not conflict with the management responsibilities of water suppliers and others.*

Development of the draft 3rd edition of the WHO guidelines for drinking water quality continued under a major work programme throughout the period 2000–2004. During that period the ideas about promoting a much more preventative risk based approach to water quality management began to be crystallised under the term "*Water Safety Plan*" (Figure 1.2).

In early 2004 IWA sponsored a further Bonn workshop with the idea of developing the themes and conclusions of the first workshop, in conjunction with the emerging water safety plan concept of WHO. This led to the development of The Bonn Charter for Safe Drinking Water (see Figure 1.3). More details are given in Appendix 1.

Both the 3rd edition of the WHO guidelines for drinking water quality and the IWA Bonn Charter for Safe Drinking Water were formally launched together in 2004. A leaflet jointly produced by WHO and IWA at that time reaffirmed the basic principles necessary to assure the safety of drinking water in the 21st century. It reminded us that:

> "*Maintenance of drinking water quality is of concern in developing and developed countries worldwide. Outbreaks of waterborne disease affecting larger or smaller populations are reported from countries at*

all levels of development and from all types of water supply. Evidence indicates that inadequate drinking water safety also contributes significantly to endemic disease. Drinking water quality has a considerable impact on health and there is a need to establish effective means of reducing the global burden of waterborne disease. There has been a growing consensus among water and health professionals about the need for a more consistent framework to be developed within which the quality of drinking water can be assured."

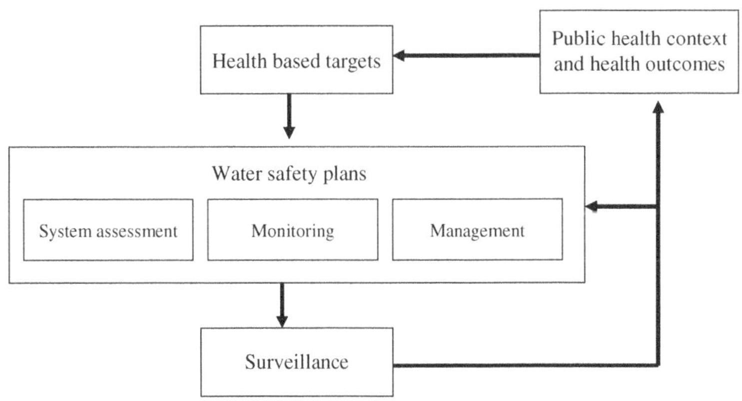

Figure 1.2 WHO framework for safe drinking water

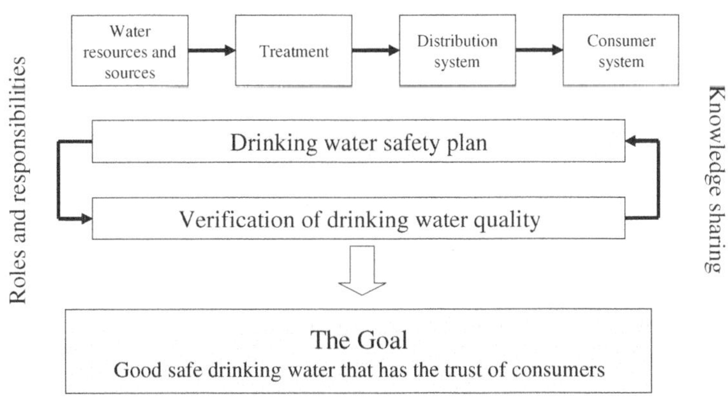

Figure 1.3 The Bonn Charter for Safe Drinking Water - summary approach

This was reinforced by the statement in the Bonn charter:

"The reliable supply of good, safe drinking water is fundamental to a healthy community and to its economic development. Its delivery requires a comprehensive understanding of contamination risks and effective control of those risks. It also requires that robust quality standards be in place and that systems be implemented to verify that safe water is produced."

1.3 WATER QUALITY OBJECTIVES FOR WATER SAFETY PLANS

By definition, the primary goal of a water safety plan is to ensure that the water delivered to consumers protects public health and is safe to drink at all times. However, in moving towards such goals, the specific standards or health targets applied may legitimately vary from country to country and over time, as recognised in the continuous review process for the WHO drinking water guidelines.

But the Bonn Charter also recognises that to meet the goal of "*Good safe drinking water which has the trust of consumers*" the water supplied has to meet objectives which go beyond simply compliance with vital health and statutory standards. It reminds us that there are three related fundamental objectives to which all those involved in the supply of drinking water should strive:

1. *"Access to good, safe, and reliable drinking water. This is one of the most basic needs of human society. In many areas water quality may already be high and continuing to improve. In others, where waterborne disease or other quality deficiencies are still prevalent, the basic provision of safe and good supplies is vital;*
2. *Water that is not just safe to drink but considered of good aesthetic quality by consumers;*
3. *Water supplies in which consumers have confidence"*

Water safety plans can address all these aspects, although in implementing such plans the priority must first and always be given to public health protection. It is for water suppliers in conjunction with the relevant authorities and their consumers to determine the goals of their water safety plan. However, these can be considered under three main headings as identified in the Bonn Charter.

1.3.1 Health and statutory standards

It is the responsibility of Government and health authorities to set standards or targets for drinking water quality which protect public health. The pre-eminent source of scientific and medical advice available to Governments globally for this purpose is the WHO Guidelines for drinking water quality (WHO 2011). In setting standards the Bonn Charter recommends that Governments should:

> *"Ensure that the process of setting quality- and health-related standards is transparent and the choices made with regard to the level of risk considered appropriate fully disclosed".*

It also suggests that "*the number of parameters covered by statutory output standards should be kept to a minimum but with*:

- *A core set of parameters tailored to meet local needs.*
- *The use of suitable "indicator" microbiological and chemical parameters, where scientifically valid.*
- *Appropriate monitoring and reporting systems.*

1.3.2 Consumer acceptability

Although the highest priority for consumers is the safety of their drinking water they also want it to taste and look good and not adversely impact on other legitimate household uses such as cooking and bathing. Equally importantly, if consumers do not like the taste or appearance of water, will they believe that the water is free of other substances that they cannot detect? In some countries statutory quality standards or consumer service levels are set for water acceptability. In other countries this is seen as a matter for the utilities themselves.

The acceptability of drinking water to consumers can vary considerably and depends on many different local factors. The causes can be complex and derive from problems in the catchment, treatment plant, distribution network, and/or consumer's own property but can lead to issues with:

- Taste and odours.
- Appearance/discolouration.
- Hardness or mineral content.
- Invertebrate animals.
- No water or low/intermittent pressure.

Importantly, it has to be recognised that the extent to which water is acceptable to consumers is often related to a change or variation in some aspect rather than the absolute level. For example, chlorine is often used as important network disinfectant residual but many consumers do not like the taste. Nevertheless, it is common experience that consumers can become used to a certain level of chlorine residual provided it does not vary unduly. However, if the level of chlorine varies considerably then this is much more noticeable and is often regarded as unacceptable by consumers. Similarly with the mineral content or hardness of water, where variation in levels can be very noticeable to consumers. More details of consumer acceptability and how it can be assessed and improved through a water safety plan approach are given in Chapter 7.

1.3.3 Quality reliability and consumer trust

Consumer trust depends not only on the consistent delivery of good safe water but also how consumers perceive its quality. The trust that consumers have in their water is impacted by a range of complex factors including:

- Consumer communication and information provision.
- Transparency of water quality information.
- Personal experience of consumers with their water supply.
- Media coverage.
- Comments by influencers of public opinion such as politicians or health experts.

However, one of the key factors is often the reliability of water quality. Water quality incidents, however small, which impact on the consumer, can significantly affect the trust that a consumer has in both the water supply and water supplier. Thus the more reliable the water quality the greater the consumer trust is likely to be. Equally, if an incident does occur, a utility's reputation with its customers, which often takes years to establish, can be lost in a matter of hours/days if handled badly. More information on managing incidents and emergencies is given in Chapter 3.

To track consumer trust, many utilities use various forms of consumer survey. Such surveys have even more value if repeated at regular intervals so that the trend in consumer views can be assessed over time as well as at a specific time (see Chapter 7).

1.4 BENEFITS OF A WSP AND RESOURCES NECESSARY FOR IMPLEMENTATION

1.4.1 Introduction

Identifying the expected benefits of a water safety plan and the potential resources that are necessary for implementation is usually a pre-requisite to:

- Securing the support of senior management.
- Ensuring that sufficient resources are made available.
- Allowing more targeted and efficient implementation.

There is a complex relationship between the benefits of, and resources needed for, WSP implementation, and the actual situation will vary considerably both from country to country and between utilities. Before implementation of a WSP it is therefore important to clearly understand the expected benefits and the associated resource needs so that priorities can be established. But long term water quality improvements which protect public health should never be neglected at the expense of short term cost savings.

1.4.2 Benefits of a WSP

The underpinning driver for a water safety plan is a clear recognition that simply relying on "*end of pipe*" water quality monitoring to manage drinking water quality is invariably too little and too late. However, the experience of water suppliers shows that there can be a wide range of benefits from implementing a WSP which can include:

1.4.2.1 Avoiding incidents which impact public health

There is a wealth of documented case studies where people have become ill or even died as a result of a preventable water quality incident (Hrudey and Hrudey 2004) in both developed as well as developing nations. If this occurs, the impact on consumers, the wider community and the water supplier can be enormous. Implementing a WSP preventative approach can significantly reduce the risk of such events occurring which can often arise through a misguided sense of complacency that "*such events could not happen here*".

1.4.2.2 Improved compliance with regulatory and other requirements

Implementing a WSP is likely to improve and/or help to maintain high levels of compliance with regulatory standards. Additionally, it can help meet other water quality standards or service level requirements such as those for consumer acceptability. In some countries water industry regulators are beginning to endorse, promote or enforce the use of WSPs. The support of regulatory agencies for WSPs is important and can offer many advantages to water suppliers. However, at the same time suppliers should avoid implementing WSPs in a token way and thus creating a perception among staff that this is being done just "*because the regulator tells us to*".

1.4.2.3 Improved consumer trust

The trust that consumers have in their water supplier is likely to be improved by the better and more reliable water quality that results from implementing a WSP. There are also likely to be a range of other related consumer service improvements e.g. a reduction in the number of water quality complaints and improvements in the response to complaints and incidents. Additionally, if communicated effectively, the development of a WSP can help to demonstrate to consumers that their utility is fully committed to provision of safe, wholesome drinking water, thus helping to build trust.

1.4.2.4 Improved confidence of key stakeholders

Improved water quality management will improve confidence in the supplier by key external stakeholders such as regulators, health authorities and municipal authorities. This in turn will help to build trust with consumers and consumer groups. WSPs may also provide a framework or basis for regular, effective communication with stakeholders such as catchment authorities or plumbers.

1.4.2.5 Cost effectiveness and investment planning

A risk based approach to water quality management should improve the cost effectiveness of the supplier's operation. This results from the considerably improved understanding and thus better control of the supply system. It can often be possible to reduce unnecessary end product monitoring and/or on site monitoring, thus again reducing operational costs.

Proactively managing risk though a WSP should also help to avoid the often significant "*costs of water quality failure*" since WSP costs are likely to be lower over the long term than continually trying to manage problems after they have occurred. Many man hours are often wasted in reactively managing such problems, and efficiency can be improved by adopting a more proactive preventative approach. Water quality incidents or failures can also have significant financial, legal and other implications for water suppliers. In a worst case even the supplier's "*licence to operate*" can be compromised. Proper implementation of a WSP should significantly reduce the risk of water quality failure, and improve the speed of recovery after incidents, thus minimising these costs.

Furthermore, a WSP approach can also help to improve targeting of capital investment by allowing better decision making on when and where to prioritise investment in water quality (see Chapter 8). Indeed in some countries support for capital investment programmes and/or consumer charges is dependent on, or can be better justified by, a risk based water quality management system. For some developing countries, support from external donor agencies may be facilitated by data provided by a WSP.

1.4.2.6 Improved staff commitment

It is often the experience of suppliers that engaging staff in a water safety plan process actually improves their commitment to the organisation and service to consumers. It also increases awareness of risk and the importance and pride that goes with protecting public health through provision of a safe and wholesome water supply. In some cases it also provides the intangible "*peace of mind*" feeling where staff know that they are doing the best they can to protect consumers and avoid major events. Wanting to provide clean, safe water to friends, family and fellow citizens can often be an important factor in staff motivation. This can be particularly important in small communities where managers feel personally responsible and are highly visible to the rest of the community.

1.4.2.7 Competition with peers

Most water utilities, whether private or public, strive to be the best and like to benchmark themselves against peers, either nationally or internationally, in the adoption of recognised best practice. Since WSPs are now regarded as best practice for water quality management their adoption should be seen as part of "*being the best*" and this in turn can help to develop a sense of pride among staff. For commercial organisations the adoption of WSPs may also help in marketing expertise or gaining new contracts.

1.4.3 The resources necessary for WSP implementation

1.4.3.1 Introduction

The resources necessary for WSP implementation will vary considerably depending on factors such as number of employees, size of asset base, service area, regulatory requirements and systems already in place. Many current work practices may act as the "*building blocks*" of a WSP and thus will reduce the amount of additional resources needed. Such building blocks may include ISO quality procedures, standardised operating manuals, maintenance schedules, monitoring programmes, staff training and stakeholder engagement programmes. However, in the end WSPs need to be developed and prioritised with the resources available. Significant improvements may often be made with little additional resource.

Thus, before starting a WSP project, it may be beneficial to undertake a gap analysis exercise to determine what aspects of the WSP are already being done (Figure 1.4). This may be helped by completing a pilot project (see Chapter 2.4.3).

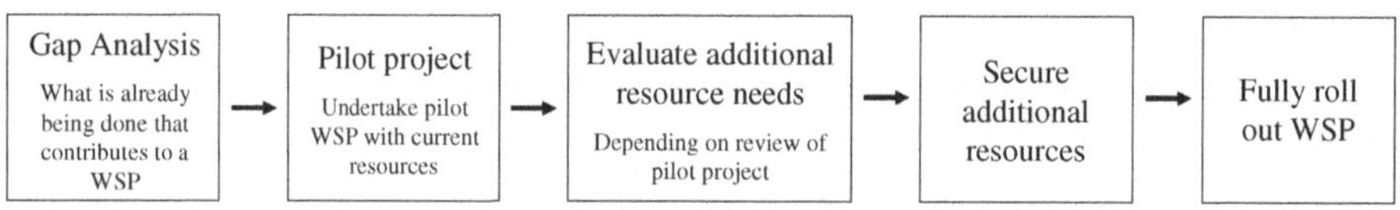

Figure 1.4 Determining resource need for WSPs

It is important that a realistic timescale for implementation of a WSP project should be identified, but this should not be underestimated, particularly in those organisations that have few existing "*building blocks*". It should also be emphasised that the WSP will never be complete and will need to be reviewed on a regular basis (e.g. annually), or as situations change. Adequate time should therefore be allocated both for initial set up and ongoing reviews.

There are three main costs associated with implementing a WSP. However, whilst the costs can be significant, particularly in larger or more complex supply systems, this should not be seen as a barrier to implementation. WSPs are scalable, and it is better to do what you can with what you have, which can still yield big improvements, rather than thinking "*we don't have the money so we won't do it*". Enthusiasm and commitment of managers and staff can be as important as lots of resource without commitment. This approach can be summarised as "*Think big, start small, scale up*" (WHO and IWA 2010).

1.4.3.2 Initial set up costs

The resources necessary to set up a WSP will vary but need to be considered under categories such as financial, staff time, materials and knowledge gathering.

Financial resources will include not only the costs of the staff involved but also such things as data gathering and assessment, establishment of IT and other management systems, attendance at training courses and internal communication. The costs of short term external consultancy support may also need to be included.

Staff time will need to be made available for WSP development, either as part of their current role, or wholly dedicated to WSP development. Whichever is the case, WSP duties should not just be added on to already busy work schedules but additional staff time made available. For example, consideration needs to be given to:

- Who will be primarily responsible for WSP production?
- Who else will need to be involved?
- Once the plans have been developed who will be responsible for update and review?
- Is there sufficient resource within the current workforce?
- Will more staff or consultants need to be employed?

Consideration should also be given to any additional materials needed to develop the WSP as well as mitigate any risks which are identified. For example:

- How will the WSP be documented and stored?
- Will any other physical materials be needed in WSP development (IT equipment, office space, consumables, documentation etc.)?
- How will materials for necessary mitigation actions be secured?

One of the main resources that the WSP itself will need is information and knowledge that will allow successful identification of hazards and assessment of the risks. Key members of the organisation who possess this knowledge should be made available for WSP implementation. Where such knowledge is not held within the organisation the help of external organisations should be sought or efforts be made to gain this knowledge.

1.4.3.3 WSP maintenance costs

Once the WSP has been established it is of vital importance that it is properly maintained, information and documentation updated, and risks regularly reviewed. The resources will not be as large as initial set up but still needs to be budgeted for and actively managed.

1.4.3.4 Operational and capital improvement programmes

Implementation of a WSP is likely to identify a need for both operational and capital improvement programmes. Depending on the utility, these might be substantial and will need to be phased over a period of time using appropriate risk prioritisation mechanisms.

REFERENCES

Fewtrell, L. and Bartram J. (2001) Water Quality: Guidelines, Standards & Health: Assessment of Risk and Risk Management for Water-Related Infectious Disease*(s)*, IWA Publishing, London.

Gunnarsdottir, M. J. and Gissurarson, L. R. (2008) HACCP and water safety plans in Icelandic water supply: Preliminary evaluation of experience, *Journal of Water and Health*, **6** (3), 377–382.

Havelaar, A. H., (1994) Application of HACCP to drinking water supply, *Food Control*, **5**(3), 145–152.

Horton, R. (1898) The chemical examination of water, *Instn. San. Engrs.*, 24–35.

Hrudey, S. E. and Hrudey, E. J. (2004) Safe Drinking Water: Lessons from Recent Outbreaks in Affluent Nations, IWA Publishing, London.

IWA (2004) The Bonn Charter for Safe Drinking Water, IWA Publishing, London.

National Health and Medical Research Council (NHMRC) (1996) Australian Drinking Water Guidelines (Online), *Available: http://www.nhmrc.gov.au/publications*, NHMRC, Canberra, Australia.

National Health and Medical Research Council (NHMRC) (2004) Australian Drinking Water Guidelines (Online), *Available: http://www.nhmrc.gov.au/publications*, NHMRC, Canberra, Australia.

Prüss-Üstün, A. and Corvalán, C (2006) Preventing disease through healthy environments: towards an estimate of the environmental burden of disease.

United Nations (2010) the Millennium Development Goals Report 2010.

Van Schothorst, M (2004) A simple guide to understanding and applying the hazard analysis critical control point concept, 3rd edition, International Life Sciences Institute, Brussels, Belgium.

World Health Organisation (1970) European Standards for Drinking Water, WHO, Geneva.

World Health Organisation (1976) Surveillance of Drinking Water Quality, WHO, Geneva.

World Health Organisation (2004) Guidelines for Drinking Water Quality, 3rd Edition, WHO, Geneva.

World Health Organisation and International Water Association (2010) A road map to support country level implementation of Water Safety Plans.

World Health Organisation (2011) World Health Assembly Resolution 64/24 on drinking water, sanitation and health.

World Health Organisation (2011b) Guidelines for Drinking Water Quality, 4th Edition, WHO, Geneva.

Chapter 2

Developing a supportive organisational culture

Corinna Summerill, Yahya Omar and Simon Pollard

2.1 THE IMPORTANCE OF ORGANISATIONAL CULTURE AND COMMITMENT FOR WSP IMPLEMENTATION

2.1.1 Introduction

The implementation of WSPs within a water utility has much in common with the adoption of any business implementation project. Implementation managers want to secure the organisational improvements and outcomes that WSPs offer, based on a more proactive, preventative risk management approach, but this will often need to be achieved in parallel with other change projects running within the utility. However, the culture of an organisation can have a major influence on the effectiveness with which new business practices are adopted, even though the nature of an organisation's culture may not be immediately obvious to its members. Successful WSP implementation therefore requires an understanding of the existing culture of the organisation and its ability to create and sustain a compelling case for the adoption of WSPs, whilst at the same time recognising wider workload and other pressures.

The process of implementing any new business process needs to be tailored to the individual utility. Whether this involves a complete redesign or simply greater formalisation of existing risk management processes, it needs a strong, respected internal champion, and the backing of Executive (Board level) authority to see it through. This is rarely straightforward because effecting change is by its very nature challenging. New systems and approaches take time to learn, and there are many competing pressures for utilities and their managers.

This Chapter seeks to provide guidance on securing the necessary managerial and organisational commitment for the implementation of WSPs based on both general management literature and specific research with water utilities. It emphasises the importance of risk management leadership and organisational culture for the effective implementation of WSPs. Most importantly, and because WSPs offer an important opportunity for a cultural shift towards preventative risk management, it draws attention to those factors that are believed to be instrumental to a sustained change of approach.

2.1.2 What is organisational culture?

As with most issues concerning risk management, the ways that both individuals and organisations behave are important because people have different individual approaches to dealing with risk, as do the organisations in which they work. The organisational culture of a water utility in relation to water quality may be influenced, among other factors, by its professional make up and history, governance model, history of previous water quality incidents and the importance with which its public health protection role is seen within its corporate strategy.

Essentially, culture can be defined as the attitudes, experiences, norms, beliefs and values of an organisation, or put more simply *"the way things get done around here"* (Deal and Kennedy 1983). Practical clues about organisational cultures are evident in the things we see, hear and feel when we enter the places where we work, for example our commitments to quality and safety, the degree of professionalism, and approaches to professional and personal development.

Therefore, attempts to shift behaviours in favour of a more preventative approach to managing water quality risks from catchment to consumer, must also consider the existing basic assumptions and culture of the organisation. It is also important to note that different sub-cultures often exist between different teams and departments of an organisation (e.g. engineers and chemists; theorists and pragmatists; strategists and tacticians), and these may even contradict one another (Smircich and Calas 1987). Some cultural aspects will work well for one organisation but not another. There is no *"right or wrong*"' overall culture. There are, however, a number of features that are common to all organisations which manage risk successfully.

2.1.3 How does culture influence WSP adoption?

A WSP aims to maintain consistently high levels of water quality but as noted above, the culture of an organisation can act as a barrier or filter to the adoption of such new practices (Johnson 1992). If the basic assumptions of the organisation do not support the aims of a new practice it may not be implemented effectively, nor the benefits sustained in the longer term.

An organisation that already has a strong risk management culture for other enterprise activities such as health and safety or operational reliability is more likely to implement WSPs effectively, simply because a risk management culture is already established. This will be further enhanced if the organisation understands and promotes its crucial public health protection role. Conversely, if the WSP approach is adopted only because it is perceived as a legal or regulatory requirement this may result in so called "*tokenism*" where the practice is implemented to please someone else and the true benefit not realised (Hamilton et al. 2006).

Those implementing WSPs aim to secure a long-lasting change of approach which engages all those involved with water quality. However, statements of intent by managers on their own are insufficient to modify employee behaviour (Fernandez-Muniz et al. 2007). Managers influence the risk culture of employees in two ways:

- Indirectly through support and funding for implementation and development of suitable management systems.
- Directly through positive attitudes and behaviours. For example, if initial senior management support for a project is not actively demonstrated through subsequent management behaviour it may simply promote a *"badge on the wall"* mentality that can lead to complacency.

Thus, whilst a high level of organisational risk management culture is not a pre-requisite for WSP adoption, it is essential that it is fully considered as part of the implementation process. Equally it may be that implementation of a WSP can be the first step in helping to improve the overall risk culture of a water supplier.

2.1.4 Managing water quality as part of overall business risk management

Many utilities now have organisation-wide risk management processes in place. In response to revised codes of good corporate governance promoted in the late 1990s in the private and public sector, many utilities put in place formal structures for capturing, assessing and managing their so-called "*enterprise risks*". Initially focussed on financial risk, enterprise risk management now incorporates health and safety hazards, reputational issues, potential environmental liabilities and a host of other threats that might impact on the strategic objectives of a utility. "*Best in class*" utilities have established chief risk officer (CRO) roles, risk governance arrangements and formal risk registers, and have empowered their workforces to understand and maintain a managed risk environment. Critically, this means balancing the need of the job with the hazards inherent to it; being able to take measured risks and seize opportunities for innovation, but without exposing the organisation or its stakeholders and customers to unacceptable exposures. For utilities with pre-existent risk management arrangements, there is the opportunity to manage water quality as part of enterprise risk management. This may afford the opportunity of tapping into wider risk expertise in the organisation, ensuring regular upward reporting to the Executive, and securing maximum reputational benefit of a well-implemented WSP.

2.2 CHARACTERISTICS OF SUCCESSFUL RISK MANAGEMENT ORGANISATIONS

2.2.1 Lessons from other sectors

Water utilities can learn useful lessons about risk management from organisations in other sectors that manage risk well and that are driven by a desire for high reliability in their operations (Hrudey et al. 2006; Pollard et al. 2007; 2009). Such high reliability organisations (HROs) have demonstrated that a high level of organisational awareness of hazard and risk is vital to effective risk management.

HROs are those which have relatively few incidents or accidents compared to the nature of their high risk operations, for example nuclear power stations, navy submarines and air traffic control centres. Leaders within HROs recognise that human error occurs, but also that human variability and an ability to adapt to changing events is an important safeguard. These organisations focus on the system as a whole seeking to remove risk causing factors through (Reason 1998):

- Establishing an effective reporting culture.
- Analysing in detail the occurrence of incidents and near misses to uncover any common themes or causes of failure.
- Striving to imagine new scenarios that could occur and protect against these.

These features contribute to developing a high reliability organisation i.e. one that has a preoccupation with the possibility of failure and its root causes (Weick and Sutcliffe 2006). Leaders are instrumental in creating this risk aware culture and leader influence has a significant effect on individual employee behaviour (Clarke and Ward 2006).

2.2.2 Criteria for successful WSP adoption

Experiences of WSP implementation are now providing valuable insights into what makes for successful uptake within utilities. Following a review of WSP guidance (Bartram et al. 2009, Godfrey and Howard 2004, IWA 2004 and NHMRC 2004), case studies (Bath et al. 2007, Gunnarsdottir and Gissurarson 2008, Tibatemwa et al. 2004 and Viljoen 2010) and the risk culture literature (Parker et al. 2006, Pollard et al. 2008, Reason 1998 and van Vuuren 2000), along with targeted research in utilities, a number of supportive cultural attributes are viewed as positively influencing WSP implementation. Utility managers can use this information to support implementation within their organisation.

- **Managerial commitment:** This is needed to support the WSP process and provide necessary resources. It requires more than just declarations of support by managers and others in positions of authority. It must also include demonstration of positive commitment to water quality through actions and behaviours.
- **Organisational commitment:** Not only is managerial commitment important but this is also required across the whole workforce. This is helped by a feeling of empowerment (see below).
- **Empowerment:** Giving ownership, involvement and recognition to staff is likely to improve motivation and commitment to successfully implement a WSP.
- **Leadership and advocacy:** Leaders within organisations are important because they are able to change attitudes and motivate others. This applies not just to senior management but also other water quality champions who are enthusiastic about WSPs, give priority to water quality and generate this feeling in others.
- **Accountability:** It is important that there is clear accountability to the consumer with utilities being responsible for their actions and willing to explain them. This is helped by creation of a feeling among staff of an obligation to do the "*right thing*" and to act in a transparent, accountable and professional manner.
- **Awareness of public health responsibility:** It is critical to avoid a sense of complacency and feeling that "*it could not happen here*" by ensuring that all staff are constantly aware of the consequence for public health of their actions or inactions.
- **Internal relationships:** Developing internal relationships is vitally important. WSPs are intended to be a holistic approach and will require co-operation from across departments, teams, and individuals at all levels. Effective teamwork and communications are essential.
- **Learning culture:** Effective WSP implementation requires an open culture that wants to learn from mistakes, past events and outside experiences. There must also be a focus on training and education, and research and development, as well as an openness to new ideas and fresh experience.
- **Open reporting culture:** This is a culture where people are comfortable to report near misses. An organisation is more open to learning if it promotes a just and "*fair blame*" culture which is honest and open in reporting.
- **Continual improvement culture:** This is important to prevent complacency and therefore "*stagnation*" of the WSP. It requires an organisation that is not content with things as they are and strives for continual improvement. For example, going beyond minimum legal requirements, gaining certifications and creating a total quality mindset.
- **Proactive:** This is linked to continual improvement culture. It means having a long term vision which recognises that the aim of the WSP is to anticipate and prevent problems occurring in the first place.
- **External relationships:** Effective WSPs need good stakeholder engagement and also to value the input of, and benefit brought by, stakeholder relationships. This needs to be coupled with an

understanding of the organisational cultures of relevant external organisations which may be different to that of the water supplier.

- **Image and reputation conscious:** Organisations which are concerned with their image and reputation often have a culture which supports WSPs. This is linked to a sense of competition and "*wanting to be seen to be the best*" amongst consumers, peer utilities and in some situations commercially.

2.3 SECURING MANAGERIAL AND ORGANISATIONAL COMMITMENT FOR WSP IMPLEMENTATION

2.3.1 Why is managerial commitment important?

Gaining managerial commitment is an essential first step in the WSP approach (Bartram et al. 2009). Without it, the project is likely to fail. Managerial commitment to WSPs will be influenced by a wide range of factors, such as existing organisational culture; regulations; political or media pressure; the experience of other utilities; current water quality; historical water quality incidents; financial resources, and other competing priorities. Where a high standard of water quality is already achieved, and where there has not been a significant water quality incident in the past, it may be more difficult to demonstrate the immediate benefit of WSPs. This does not mean that WSPs are unnecessary or unimportant but there is a risk of water quality becoming taken for granted.

It is particularly vital to emphasise the paramount importance of public health protection as an objective of the WSP process otherwise this may be lost among other legitimate drivers (Summerill et al 2010; Summerill et al., 2010b). However, managers also need to appreciate the many other benefits of WSPs which are summarised in Chapter 1. Indeed, sound risk governance is a strategic business asset in its own right and can help build improved stakeholder confidence as well as provide many other benefits. There needs to be a recognition that long term improvements to water quality management should not be compromised by short term financial pressures, and that WSPs can actually help to inform sound investment and operational planning.

2.3.2 Demonstrating managerial commitment

As described previously, managerial commitment should not be seen just as the provision of resources or formal endorsement of future policy, although these are important. Managers also need to demonstrate commitment by becoming actively involved in WSPs through a range of means such as:

- Formally endorsing a WSP policy.
- Ensuring sufficient resources, both human and financial.
- Actively encouraging workforce commitment and buy-in to WSPs.
- Avoiding WSP implementation "*tokenism*". For example, appearing to support adoption solely because it is a regulatory requirement.
- Promoting the necessary change in risk culture to support WSP implementation.
- Promoting the importance of the public health protection role in the workforce.
- Supporting necessary changes in work practices.
- Setting and monitoring clear water quality targets, an important preliminary step of the WSP.
- Identification and engagement with relevant external stakeholders.
- Provision of training, education, and succession planning.
- Taking an active interest in risk assessment and mitigation.

2.3.3 Employees wishing to secure managerial commitment

In some cases the managers of water utilities may not be fully aware of WSPs or their benefits. Thus if employees wish to secure the commitment of their managers to WSP implementation there are a number of approaches that might be considered (Godfrey & Howard 2005). These are summarised below and set out as a flowchart in Figure 2.1:

- Ensure managers are fully aware of what the WSP approach means.
- Promote the WSP benefits that will be specific to the particular needs of the organisation.
- Highlight past water quality incidents, either within the organisation or elsewhere, which could have been prevented using the WSP approach.
- Illustrate how existing policies and procedures within the utility may act as the "*building blocks*" of a WSP, for example asset management and enterprise risk management.
- Recommendation of a pilot study to demonstrate benefits.
- Tailor the approach to the current organisational priorities and leadership style.

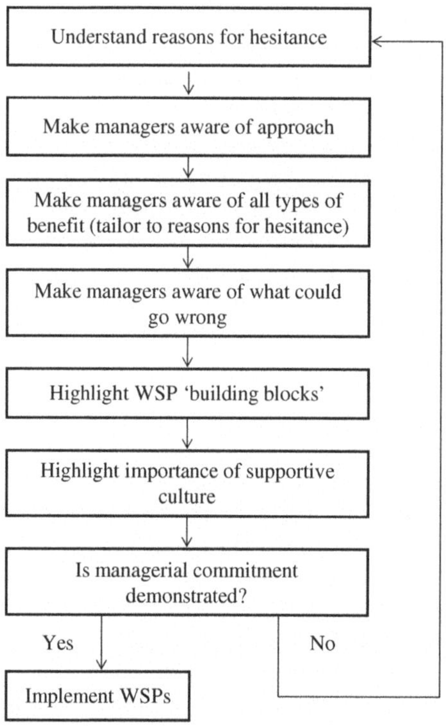

Figure 2.1 Steps to gain managerial commitment

2.3.4 Organisational commitment and motivation

Not only is managerial commitment important, but the commitment of the whole organisation is necessary for effective WSP implementation. Since any major change in practice requires commitment and motivation, managers and leaders need to understand the factors that underpin this.

2.3.4.1 *Securing staff commitment*

Staff may be committed to an organisational goal for a number of reasons. Meyer and Allen (1991) identify three key motivations, set out below in Table 2.1, with examples relating to water utilities. In recent research, more reasons related to category 1 were given for commitment to the organisation, to the individual's job role and to WSP implementation. This positive emotional attachment (actually wanting to be committed) emphasises the vital importance of creating the appropriate organisational culture during WSP development. Researchers maintain that creating conditions in which employees *want* to be committed is the most successful in creating positive behavioural change, rather than feeling they have no choice, or that they ought to be (Meyer and Herscovitch 2001).

Table 2.1 Reasons for organisational commitment

Reason	Examples from research
1. Staff want to be committed	• Employees wanted good quality water themselves, and also for friends and family who lived in the area. • Non-monetary benefits, such as celebrations of success, giving staff flexibility in work patterns and giving staff variety in their jobs. • Cultural factors such as team support, enjoyment at work, and the feeling that staff were providing a service. • Camaraderie, wanting to work well for other team members, "*pulling one's weight*". • Empowerment and provision of information. • Feeling that they were appreciated by management. • In some areas religion was a very important factor in staff commitment.
2. Staff have no choice but to be committed	• Managers felt they would be "*in the firing line*" from residents if something went wrong, and hence had a strong incentive to do the best they could. • Pay progression and opportunities for career progression. • Monetary bonuses, corporate balanced scorecard award programme. • Have worked for the company for a long time.
3. Staff ought to be committed	• Prior training and education. • Training and continuing education needed for operator licences. • Promotion.

2.3.4.2 *Securing staff motivation*

Staff can be motivated to implement goals for both emotional and material reasons. Many believe that emotional factors, i.e. that a person wants to do something, is much more important in the long term rather than simply providing material benefits, such as a bonus, if the work is carried out well. However, research suggests that although material things such as salary, benefits and job security will not motivate in their own right, an absence of these factors can result in de-motivation (Herzberg et al. 2004) so although important, they should not be considered in isolation. During interviews with water utility employees, the following factors were mentioned as motivating factors. All can be used by managers to encourage and motivate staff for WSP development:

- Empowerment, involvement and ownership in projects such as WSPs.
- Provision of awareness and education.

- Recognition for good work.
- A good working environment.
- Wanting to avoid repetition of past incidents.
- Regulations and guidelines.
- Feeling of providing a service to community.
- Feeling of being protectors of public health.
- A desire to do the right thing.
- Image: wanting to work for the "*best*", or a "*world class*" organisation.
- Competition with peers.
- A feeling of being trusted.
- Pride in work.
- Practices that "*make my job easier*".

2.3.5 Awareness of the importance of water quality and public health protection

Critical to the successful implementation of WSPs is promoting among all staff an awareness of the importance of drinking water quality and public health protection. Hrudey et al. (2006) suggested the following approaches for adoption by water suppliers:

- Water quality vigilance is actively promoted and rewarded.
- Promoting and actively maintaining an understanding of the entire water supply system, its challenges and limitations.
- Adopting effective, real-time treatment process control, based on an understanding of critical capabilities and limitations of the technology.
- Fail-safe multi-barriers are actively identified and maintained at a level appropriate to the challenges facing the system.
- "*Close calls*" are documented and used to train staff about how the system responded under stress and to identify what measures are needed to make such events less likely in the future.
- Operators, supervisors, lab personnel and management all understand that they are entrusted with protecting the public's health and are committed to honouring that responsibility.
- Operational personnel are afforded the status, training and rewards commensurate with their responsibilities as guardians of the public's health.
- Response capability and communication are improved.
- An overall continuous improvement, total quality management mentality pervades the organisation.

2.3.6 WSP awareness programmes

Research with utilities on WSP implementation regularly identifies the importance of WSP awareness and education to securing commitment. For example, in one utility where WSPs had failed, this was attributed primarily to a lack of communication and awareness. By contrast, in other utilities where WSP implementation has succeeded, awareness programmes were key to the process.

Adoption of WSPs should engage everybody in the Bonn Charter goal of providing "*good safe drinking water that has the trust of consumers*" (IWA 2004). Gaining commitment is likely to be easier if staff are first made aware of the project, what it entails and the likely benefits. Conducting a pilot project (see section

2.4.3) may help in gaining commitment. Some other aspects that may be considered in a WSP awareness programme include:

- Prioritisation: Those who are directly involved first.
- Promotion: All staff are helping to provide safe drinking water.
- What is a WSP?
- Why are we doing WSPs?
- What are the likely costs and benefits?
- How will it affect people's jobs?
- Encourage people to think about hazards and risks.
- For those who will be directly involved, follow up with training.

2.4 IMPROVING ORGANISATIONAL CULTURE FOR WSP IMPLEMENTATION

2.4.1 Developing and improving risk leadership

Even with the most effective business processes for risk management, risks do not get managed unless organisations and the individuals within them recognise and accept accountability and take appropriate actions. Creating a compelling case for change and galvanising adoption of new practice is the role of leaders. In the context of WSPs it is important to understand leadership, because leaders not only influence organisational culture but also have the power to change culture.

Leadership and management are different attributes that complement one another. Management is about coping with complexity through planning and setting targets, ensuring a plan is accomplished through controlling and problem solving. Leadership on the other hand is about driving change by setting direction, aligning resources and securing a vision by motivating and inspiring people. Leaders influence, inspire and motivate others. Acceptance of a new practice as the norm is strongly affected by the actions of leaders (Tolbert and Zucker 1996).

Leadership can thus be defined as:

> *"The process of influencing others to understand and agree about what needs to be done and how to do it, and the process of facilitating individual and collective efforts to accomplish shared objectives".*
>
> (Yukl, 2006)

Based on existing literature (Kotter 1995, Kouzes and Posner 2002) and reinforced by recent research, when implementing WSPs leaders may wish to consider the following:

- Understand that the actions of leaders are instrumental in bringing about change.
- If the WSP is believed to be of value, then it should be clearly highlighted as a priority.
- Demonstrating commitment through active involvement in the process.
- Developing a powerful "*guiding coalition*". The WSP team needs to have the necessary authority and backing to be able to implement change.
- A clear vision for the WSP, including identifying and communicating the reasons for implementation to all staff.
- Assess what is blocking effective WSP development, such as a lack of resources or poor communication and resolve these.

- Plan for, and celebrate short term wins to generate motivation. For example, promote the success of a pilot WSP project at a water treatment plant that shows the greatest improvement.
- Anchor changes in corporate culture. For example, if formal targets and recognition are important, then ensure that all necessary staff have WSP duties within their targets.
- Ensure that the WSP works for the organisation and do not blindly follow the methods used by others.

Figure 2.2 illustrates, using quotes from representatives of water utilities studied in a recent research project, how leadership problems contributed to problems in WSP implementation. For example, in one case leaders understood that communication may have been an issue, but the leader was not even present during WSP communication exercises to employees. This may have then conveyed a negative message about its relative importance to the leader. Such quotes can be aligned with Kotter's "*leadership errors*" (1995) which are identified by number in the Figure:

1. Not establishing a great enough sense of urgency.
2. Not creating a powerful enough guiding coalition.
3. Lacking vision or under-communicating the vision.
4. Not removing obstacles.
5. Not systematically planning for and creating short term wins.
6. Not anchoring changes in corporate culture.

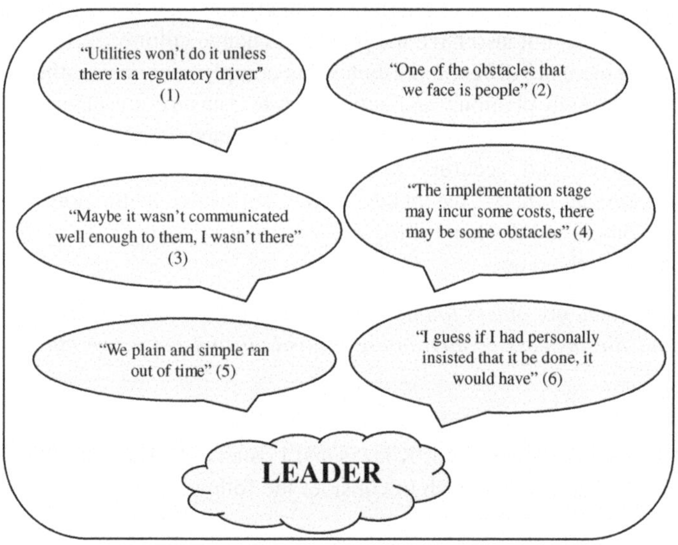

Figure 2.2 Examples of leadership errors in relation to WSP implementation

2.4.2 Improving organisational WSP culture

Evidence shows that organisational culture can be changed and improved through effective leadership. For WSPs to be implemented effectively a proactive, risk management organisational culture which values the role of public health protection highly needs to be developed. Understanding cultural barriers to, and enablers for, WSPs is an essential first step for leaders wishing to create a supportive WSP culture. Culture cannot be changed overnight, and is a long term goal which aims for sustainable implementation of WSPs. However, if done effectively, implementing WSPs may also help in modifying overall organisational culture.

Positive words and formal policy statements are insufficient on their own to modify employee behaviour (Fernandez-Muniz et al. 2007). Managers must become personally involved through day to day actions in order to demonstrate commitment and modify behaviours - demonstrating leadership rather than just management. Good risk management goes beyond simply knowing the right words to use (Flin 2003). Too often, leaders think that they can achieve low risk operation by publicly stating that this is a priority, but then leaving it to others to implement. This can inadvertently convey a false message about priorities (Hopkins 2005). There are a number of important aspects set out below which need to be considered when seeking to improve organisational risk culture.

2.4.2.1 *Mission statements*

When properly developed, these can help focus an organisation on "*what really matters, to itself as well as to its stakeholders*" (Ireland and Hitt 1992). A well developed mission statement can help generate commitment throughout the organisation, provide direction and increase performance (Forehand 2000). But many organisations do not develop them, or if they do, they fail to consider the needs of stakeholders and consumers (Ireland and Hitt 1992).

The Bonn Charter stresses that a primary goal of water suppliers should be "*good, safe drinking water that has the trust of consumers*". For most utilities there are a wide range of expectations among external stakeholders which, as well as consumers, may include governments, regulators, health agencies, and contractors. In this situation the critical public health component may be at risk of getting lost. If mission statements are a way of establishing direction and performance, neglect or downplaying of public health aspects may inadvertently create a message that devalues or de-prioritises initiatives such as WSPs. This does not mean that other utility objectives are less important, but perhaps the water safety goal requires renewed visibility. With the complex challenges facing water utilities today, it is their leaders who are responsible for making it clear to its employees what are the priorities for their organisation and, if implementing WSPs, how this fits in.

2.4.2.2 *Internal cultures*

There may be different cultures present between different departments or different professions. For example, the engineering department may have a culture which may be different to that of operational or scientific departments. They may therefore view WSPs in different ways. Here communication and understanding is needed, different sub-cultures may need to focus more on understanding each other's cultures in order to prevent these creating a barrier for WSP implementation. Effective communication is needed to promote the purpose of the WSP, so that its meaning does not get diffused by different internal cultures.

2.4.2.3 *Types of working structure*

The types of working practices may vary considerably, and different approaches may work for different organisations. For example very formal or "*mechanistic*" cultures may be highly structured, potentially bureaucratic, seek loyalty and obedience and are stable. Less formal "*organic*" organisations are less structured, usually smaller and more open to change. Such cultures may, however, create some challenges in WSP implementation. WSPs require formal procedures and practices, but this may be more difficult for informal cultures to adopt. Informal organic organisations may find it easier to change work practices and culture than mechanistic ones (Burns and Stalker, 1994). More formal, target dependent organisations may find change more challenging and view WSPs as another process to be followed, and may neglect instilling the WSP culture within the organisation.

2.4.2.4 Blockers to WSP development

Research among utilities identified some potential features which might be a block to WSP implementation including lack of resources and skills, as well as more cultural features such as lack of awareness; uncertainty; lack of recognition; complacency; poor internal relationships and communication; competing priorities; and contrasting internal cultures. Leaders within utilities should therefore be aware that challenges to WSP implementation may not purely be physical such as the frequently mentioned lack of resources or time. Deeper cultural reasons may come into play, and may be more important than resource provision. Poor communication within the organisation; uncertainty over how to implement the approach and competing priorities; conflicting internal cultures between departments; complacency over quality and a lack of recognition for staff working on such projects will have potentially major impact. Leadership directly influences culture and therefore it is only through effective leadership that these blocking features can be removed.

2.4.2.5 External implementers

In many cases WSPs are promoted or encouraged by other external organisations such as aid agencies, regulators, consultants and research institutions. It is important that such organisations acknowledge the impact organisational culture can have when seeking to influence the adoption of new practices. Otherwise if they do not seek in parallel to develop a "*WSP culture*" within the organisation, when they leave, the practice is unlikely to be sustainable and continue effectively.

2.4.2.6 WSPs help to create culture

There are many added value benefits of WSP implementation (see Chapters 1 and 8). But it should be recognised that the act of implementing WSPs will also often create positive cultural changes such as prevention of complacency; providing peace of mind and creating a focus on critical aspects.

2.4.3 Piloting WSPs

Before dedicating considerable amounts of time and resources to a full WSP project, it may be beneficial to undertake a pilot project in order to demonstrate the costs and benefits to senior management, the organisation as a whole, and other stakeholders. Pilot projects are used to put into action a programme or an idea so that issues and obstacles to successful implementation are identified and resolved quickly and to fine-tune the methods used before wider roll out. This demonstrates feasibility and advantages of the approach which in turn may lead to acceptance of the WSP method. WSP pilots can demonstrate benefits without the need for considerable resource and also provide a guide for the amount of resources, both financial and human, that might be needed for wide scale implementation (Figure 2.3). The following suggestions are made to those wishing to undertake a pilot WSP project:

- Make participants aware from the outset that WSP is not a one-off exercise, but rather a permanent and continuous process.
- WSPs need to be an integral and internalised part of the water supply process rather than additional interventions.

- Stakeholders need to be involved in the process in order to maintain the catchment to consumer aspect of WSP; such relationships can be tested during the pilot stage.
- Even though guidance states it is essential to train managers, non managerial employees also must be made aware of what the WSP process is and what it stands for. It is important that this is done from the beginning of the process.
- The planning stage is very important for a pilot project. Those involved need to consider: What are the reasons for doing it? What will happen when the pilot stage is over?
- Small utilities with only one supply system may perform the pilot on an individual section of the supply chain (e.g. treatment or distribution) rather than one whole supply (catchment to consumer).
- WSPs are case specific and need to be developed in context with existing working practices using appropriate tools developed locally or from within each specific organisation. One should expect to amend initial plans following a pilot.
- For the pilot project, focus on an area where rapid benefits will be evident, the so called "*quick wins*". This will help in motivation for a wider scale project.

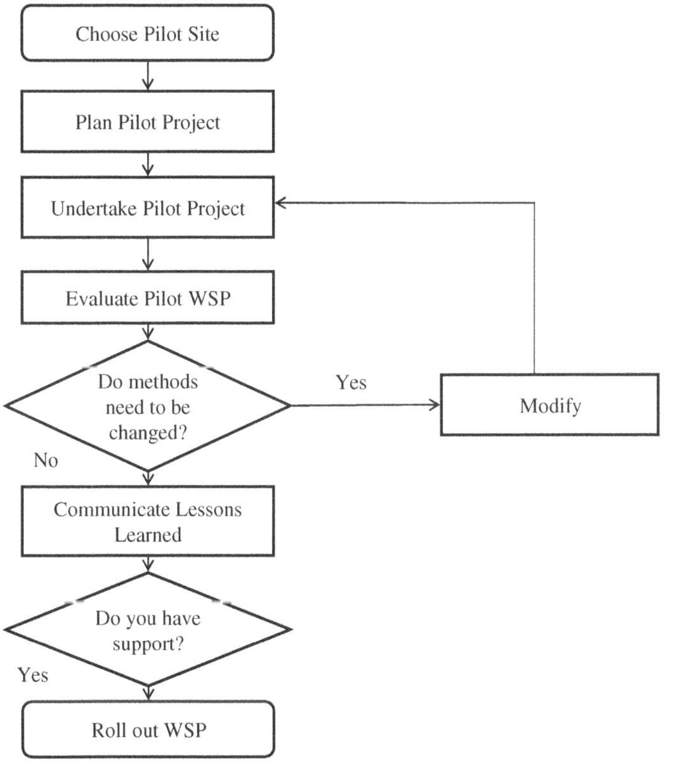

Figure 2.3 WSP pilot project flow diagram

Awareness needs to be raised that not all the benefits will be seen during the pilot; many benefits will accrue in the long term and may differ from situation to situation. For example, where water quality is currently poor, the pilot project may see an increase in quality and decrease in risk; but where water quality is currently good, benefits may be various, such as operational, improved knowledge, and better procedures. Pilots should run for meaningful periods of twelve months or more (Davison et al. 2005) and include an evaluation phase to ensure the monitoring tools have been used and appropriate actions taken. As part of the evaluation, it would be important to evaluate the viability of the supporting programmes. Given the limited time frame for a pilot project, care needs to be taken in evaluating the successful implementation of such projects. Commentators on pilot project management offer some tips such as using your best team; always documenting lessons learned; developing a mission statement with a goal and using workshops to gather input from all stakeholders.

REFERENCES

Bartram, J., Corrales, L., Davison, A., Deere, D., Drury, D., Gordon, B., Howard, G., Rinehold, A. and Stevens, M. (2009) Water Safety Plan Manual: Step-by-step risk management for drinking-water suppliers, World Health Organisation, Geneva.

Bath, A., Bowman, M. and Walker, R. (2007) Development of Water Safety Plans in the Provision of Safe Drinking Water in Western Australia. *Presented at 'OzWater' conference, Australian Water Association 2007.*

Burns, T. and Stalker, G. M. (1994) The Management of Innovation, Oxford University Press, Oxford.

Clarke, S. and Ward, K. (2006) The role of leader influence tactics and safety climate in engaging employees' safety participation, *Risk Analysis*, **26**(5), 1175–1185.

Davison, A., Howard, G., Stevens, M., Callan, P., Fewtrell, L., Deere, D. and Bartram, J. (2005) Water Safety Plans Managing drinking-water quality from catchment to consumer, World Health Organisation, Geneva.

Deal, T. E. and Kennedy, A. A. (1983) Culture - A new look through old lenses, *Journal of Applied Behavioural Science*, **19**(4), 498–505.

Fernandez-Muniz, B., Montes-Peon, J. M. and Vazquez-Ordas, C. J. (2007) Safety culture: Analysis of the causal relationships between its key dimensions, *Journal of Safety Research*, **38**(6), pp. 627–641.

Flin, R. (2003) Danger, men at work: Management influence on safety, *Human Factors and Ergonomics in Manufacturing*, **13**(4), 261–268.

Forehand, A. (2000) Mission and organisational performance in the healthcare industry, *Journal of Healthcare Management*, **45**(4), 267–277.

Godfrey, S. and Howard, G. (2004) Water Safety Plans (WSP) for Urban Piped Water Supplies in Developing Countries, Water, Engineering and Development Centre, Loughborough University, UK.

Gunnarsdottir, M. J. and Gissurarson, L. R. (2008) HACCP and water safety plans in Icelandic water supply: Preliminary evaluation of experience, *Journal of Water and Health*, **6**(3), 377–382.

Hamilton, P. D., Gale, P. and Pollard, S. J. T. (2006) A commentary on recent water safety initiatives in the context of water utility risk management, *Environment international*, **32**(8), 958–966.

Herzberg, F., Mausner, B. and Snyderman B.B. (2004) The Motivation to Work, 7th ed., Transaction Publishers, New Jersey.

Hopkins, A. (2005) Safety, Culture and Risk. The Organisational Causes of Disasters, CCH, Sydney, Australia.

Hrudey, S. E. and Hrudey, E. J. (2004) Safe Drinking Water: Lessons from Recent Outbreaks in Affluent Nations, IWA Publishing, London.

Hrudey, S. E., Hrudey, E. J. and Pollard, S. J. T. (2006) Risk management for assuring safe drinking water, *Environment International*, **32**(8), 948–957.

Ireland, R. D. and Hitt, M. A. (1992) Mission Statements: Importance, Challenge and Recommendations for Development, *Business Horizons*, May-June 1992.

IWA (2004) The Bonn Charter for Safe Drinking Water, IWA Publishing, London.

Johnson, G. (1992) Managing Strategic Change - Strategy, Culture and Action, *Long range planning*, **25**(1), 28–36.

Kotter, J. P. (1990) A Force for Change: How Leadership Differs from Management, The Free Press, New York.

Kotter, J. P. (1995) Leading Change - Why Transformation Efforts Fail, *Harvard business review*, **73**(2), 59–67.

Kouzes, J. and Posner, B. (2002) The Leadership Challenge, 3rd ed., Jossey-Bass, San Francisco, CA.

MacGillivray, B.H. and Pollard, S.J.T. (2008) What can water utilities do to improve risk management within their business functions? An improved tool and application of process benchmarking, *Environ. Intl.* **34**, 1120–1131.

Meyer, J. P. and Allen, N. J. (1991) A three-component conceptualization of organisational commitment, *Human Resource Management Review*, **1**(1), 61–89.

Meyer, J. P. and Herscovitch, L. (2001) Commitment in the workplace: toward a general model, *Human Resource Management Review*, **11**(3), 299–326.

National Health and Medical Research Council (NHMRC) (2004) *Australian Drinking Water Guidelines (Online). Available: http://www.nhmrc.gov.au/publications/synopses/eh19syn.htm Accessed on [5th July 2008],* NHMRC, Canberra, Australia.

Parker, D., Lawrie, M. and Hudson, P. (2006) A framework for understanding the development of organisational safety culture, *Safety Science*, **44**(6), 551–562.

Pollard, S., Hrudey, S. E., Hamilton, P., MacGillivray, B., Strutt, J., Sharp, J. Bradshaw, R., Leiss, W. and Godfree, A. (2007) Risk analysis strategies for credible and defensible utility decisions, Awwa Research Foundation research report 91168, Awwa Research Foundation, American Water Works Association and IWA Publishing, Denver, CO, Ref 1P-3.25C-91168-02/07-NH, 88pp.

Pollard, S. J. T., Bradshaw, R., Tranfield, D., Charrois, J. W. A., Cromar, N., Jalba, D., Hrudey, S. E., Abell, P. and Lloyd, R. (2008) Developing A Risk Management Culture – 'Mindfulness' in the International Water Utility Sector, Awwa Research Foundation, Denver, Colorado.

Pollard, S. J. T., Bradshaw, R., Tranfield, D., Charrois, J. W. A., Cromar, N., Jalba, D., Hrudey, S. E., Abell, P. and Lloyd, R. (2009) Developing a risk management culture – 'mindfulness' in the international water utility sector (TC3184), Water Research Foundation, Denver, USA, ISBN 978-1-60573-045-5, 117 pp.

Reason, J. (1998) Achieving a safe culture: Theory and practice, *Work and Stress*, **12**(3), 293–306.

Smircich, L. and Calas, M. B. (1987) Organisational culture: A critical assessment., in Jablin, F. M., Putnam, L. L., Roberts, K. H., et al (eds.) Handbook of organisational communication: An interdisciplinary perspective, Thousand Oaks, CA, US., 228–263.

Summerill, C., Smith, J., Webster, J. and Pollard, S. (2010) An international review of the challenges associated with securing 'buy-in' for water safety plans within providers of drinking water supplies, *Journal of Water and Health*, **8**(2), 387–398.

Summerill, C., Pollard, S. J. T. and Smith, J. A. (2010b) The role of organizational culture and leadership in water safety plan implementation for improved risk management, *Science of the Total Environment*, **408**, 4319–4327.

Tibatemwa, S. M., Nabasirye, L. and Godfrey, S. (2004) Implementing water safety plans: Experiences from Uganda, Conference paper, 30th WEDC International Conference, Vientiane, Lao PDR, 2004.

Tolbert, P. and Zucker, L. (1996) The institutionalization of institutional theory, in Clegg, S., Hardy, C. and Nord, W. (eds.) Handbook of organisation studies, Oaks, CA.

Van Vuuren, W. (2000) Cultural influences on risks and risk management: six case studies, *Safety Science*, **34**, 31–45.

Viljoen, F. C. (2010) The World Health Organisation's water safety plan is much more than just and integrated drinking water quality management plan, *Water Science & Technology*, **61**(1), 173–179.

Weick, K. E. and Sutcliffe, K. M. (2006) Mindfulness and the quality of organisational attention, *Organisation Science*, **17**(4), 514–524.

Yukl, G. (2006) Leadership in Organisations, 6th ed., Pearson Education.

Chapter 3

Operational policies to support water safety plans

Bob Breach

3.1 INTRODUCTION

Water safety plans are at the heart of effective drinking water quality management. However, they should not be seen in isolation from the many other policies and procedures that water utilities need to develop for a range of purposes. Indeed, many utilities developing a WSP for the first time will recognise that they already have many elements in place which can integrate well with the WSP. This chapter summarises the key principles of risk management for WSPs and then describes some of the typical water utility policies and procedures which are necessary to complement and support them.

3.2 RISK MANAGEMENT PRINCIPLES FOR WSPS

3.2.1 Introduction

Implementation of WSPs, particularly in larger utilities, can be a complex process, require a variety of resources and take time. Therefore before establishing such a system it is important to be clear on the most cost effective approach to meet local priorities and needs. This will need to take into account a number of factors, including the current water quality performance, the water quality improvement goals that have been set, the complexity of the water supply system, the extent to which risk management procedures have already been put in place, and the financial and manpower resources available for implementation. However, lack of resources should not mean that the adoption of water safety plans is delayed, rather that the approach should be phased and prioritised.

There are many *hazards* and *hazardous events* from catchment to consumer that could to some degree and at some time compromise drinking water quality (see Chapters 4–7). To manage these effectively, water suppliers need an understanding of the *likelihood* that these will occur and the *consequence* if it does i.e. quantify/determine the actual *risk* to water quality. This then allows decisions to be taken on what risk control measures need to be put in place or further improved and the priority for action.

Additionally, water suppliers may wish to integrate water quality risk management with other forms of business risk management including for example health and safety, and operational reliability.

The effectiveness with which risk management is implemented depends not just on the technical competence of an organisation, but also whether the culture and attitude of managers and staff is supportive of a risk based approach. Both aspects must be addressed. More information is given in Chapter 2.

This Section provides an overview of typical risk management approaches for water quality management which is summarised below in Figure 3.1. Much more specific detail can be found in the WHO Guidelines for Drinking Water Quality (WHO 2011), Water Safety Plan Manual (Bartram et al 2009) and other references (Davison et al 2005, Godfrey and Howard 2004).

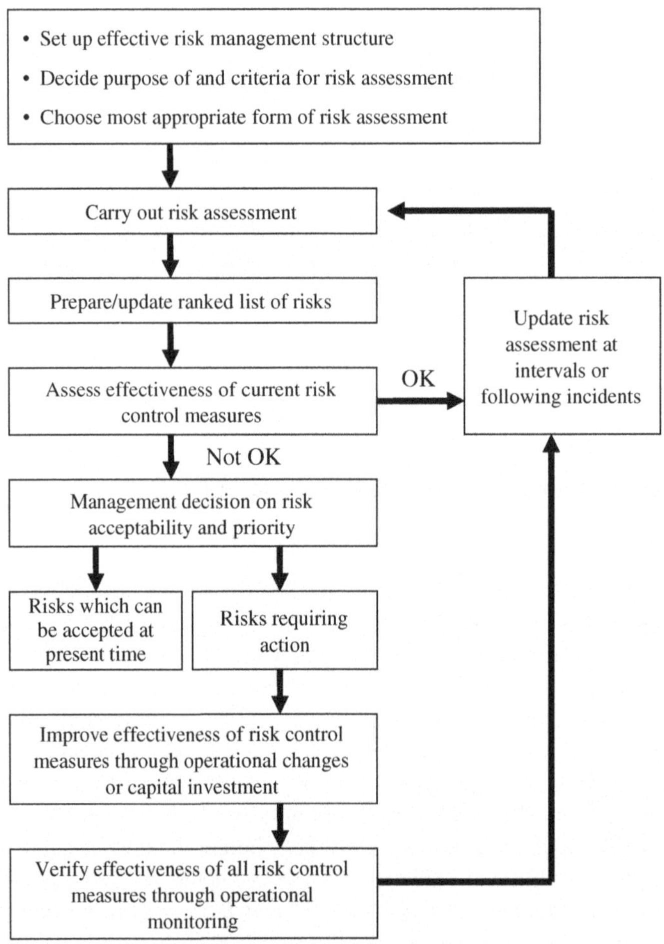

Figure 3.1 A typical approach to water quality risk management

3.2.2 Setting up a risk management structure

Successful implementation of a WSP is critically dependent on the establishment of an effective management structure. Before setting up a team to develop WSPs it is vital that this has the full and active support of senior management. This will be greatly helped by having a clear understanding of the benefits to be derived from the WSP and the costs and resources necessary for implementation (see Chapter 1). The main role of senior managers (see Chapter 2) is:

- To provide adequate resources both in terms of staff time and necessary financial support.
- To be seen to actively support the project and lead any necessary changes in culture and working practices.

It is also important that senior management clearly defines the responsibilities of the WSP team including the decisions which the team can take and those which must be referred back to senior management. There must also be a regular reporting system to senior management which summarises progress against project milestones, actions taken and issues which require decision.

Establishment of an experienced, multidisciplinary team that has the skills and knowledge to develop and implement a WSP is also vitally important. The team must have a good understanding of all parts of the water supply system and the associated risks. The approach adopted may vary depending on whether the system is large or small, for example:

- ***Large suppliers*** with multiple or complex supply systems may require more than one WSP working group reporting to a central coordinating team.
- ***Small suppliers*** may only need one WSP team but may not have sufficient in-house resources. Thus external support may be needed.

A team leader should be appointed to drive the project and maintain focus on priority tasks. This person should have sufficient delegated authority as well as the organisational and interpersonal skills to ensure effective implementation.

The team should be responsible for development, implementation and maintenance of the WSP as part of their day-to-day roles. Involvement of operational staff will contribute to success through facilitation of ownership and implementation. Team members should collectively possess the skills required to identify and assess water quality risks and how they may be controlled. Where required skills are unavailable locally, the team leader should seek external support. The team should have individual members' responsibility clearly defined and recorded at the start of the process.

One of the key initial tasks of the WSP team is to describe and document the water supply system from catchment to consumer. Depending on the availability and accuracy of existing information on catchments, water treatment works, distribution networks and consumer systems, this may take time and resources to prepare and validate.

Establishment of a WSP is initially therefore, likely to require considerable time input, with team members spending time in the field inspecting the system and working with local operational staff to describe and document the system, identify risks, and the effectiveness of risk control measures, if any, that are already in place. Once the WSP has been established the time input for ongoing reviews and maintenance should decrease.

3.2.3 What form of risk assessment to use?

There are a number of different approaches to risk assessment ranging from simple to very complex and sophisticated. The cost, time and resource requirement increases with the degree of complexity. The types of risk assessment can be broadly categorised into two types:

- Qualitative or semi qualitative, e.g. 5x5 matrix which is by far the most common approach currently used for WSPs.
- Quantitative methods, e.g. fault tree analysis.

Before commencing the process it is therefore important for utility managers to decide on the most appropriate type of risk assessment for their organisation. In the first instance, a relatively simple approach such as the 5x5 matrix is likely to produce the most benefit. However, where justified or as the organisation becomes more experienced in risk management, other more sophisticated and/or complex forms of risk assessment might be appropriate (MacGillivray et al. 2006, Rosen et al. 2007, Hokstad et al. 2009).

It should also be noted that one of the most common difficulties that water suppliers experience when implementing WSPs using a qualitative or semi qualitative risk assessment approach is ensuring consistency of the risk score between the large numbers of people that might be involved. It is thus important to provide clear training and guidance to those involved and make sure that the WSP coordination team has a mechanism for moderating the scores to ensure that the assessment is as fair and balanced as possible across all assets and risks.

3.2.4 Criteria for action and risk acceptability

Whilst a water supplier will wish to take all practicable steps to minimise risks to water quality, particularly those impacting public health, it is unrealistic to assume that all risks, however low, can be completely avoided. Indeed many lower level risks will not be capable of being addressed, at least in the short to medium term. Thus once a list of all potential risks has been identified a process will have to be put in place to rank these risks according to priorities identified by utility management. Decisions will then have to be taken about:

- Which of the highest risks must be addressed as a matter of priority?
- Which risks can be mitigated in the short term in full or part by changes to operational procedures?
- Which risks need to be addressed in the medium term e.g. as part of a capital improvement programme?
- Which risks can be accepted for the present?

This will then allow an improvement plan to be put in place and its delivery monitored. Even if immediate measures to deal effectively with the risk cannot be secured then it might be possible to reduce the risk through short term operational or procedural improvements and this needs to be built into the improvement plan, along with a timescale for more comprehensive risk reduction measures, for example as part of capital improvement schemes.

3.2.5 Control measures

There are a wide range of potential measures that may be used to mitigate the water quality risks that have been identified as part of the WSP. Each risk should have a documented control measure that eliminates the risk or reduces it to an acceptable level. Each control measure will need to be *validated* to demonstrate that it does effectively mitigate the identified risk to the level expected, for example whether the chosen disinfection regime is sufficient to deal with the assessed pathogen risk. The documentation for each control measure should include:

- The hazard and risk that is being addressed.
- The control measure that is being used to mitigate the risk.
- The control point at which the effectiveness of the measure can be assessed.
- The critical limit that has been set to determine the effectiveness of the control measure.
- Who is responsible for operational monitoring to ensure that the critical limit is met and how often is this done.
- The corrective action that is necessary if operational monitoring shows that the critical limit has been exceeded.

Since most WSPs are likely to identify a large number of potential risks, it is often considered helpful for these to be grouped into various categories and scale of significance. Those that are most critical for ensuring public health (for example disinfection effectiveness) may be designated as a critical control point (CCP).

Typical risk mitigation measures and associated control points that might be used to address common water quality risks are described in more detail in Chapters 4–7. However, Table 3.1 below sets out some simple examples of different types of risk controls that might be used.

3.2.6 Assessing WSP effectiveness

It is important that water utilities regularly assess the effectiveness of the WSP. This is normally carried out by a regular audit to ensure that all the risks have been properly and comprehensively evaluated and adequate risk mitigation measures are in place and effective. Some utilities may carry out such audit internally using a team that is not involved in day to day operational management. However, both WHO and the Bonn Charter recommend that an independent third party verification of the effectiveness of the control systems is also put in place. The Bonn Charter also reminds us that "*Irrespective of how such verification is achieved, it should not conflict with the direct management responsibility of water suppliers and others.*"

3.2.7 WSP documentation and review

Effective documentation of the WSP is a vitally important task. In a small utility a paper based version may be possible, but in most utilities the WSP is likely to be held electronically in a spreadsheet or database. If the WSP covers a large asset base the documentation is likely to be large and subject to continual review and updating. It thus needs to be designed and held in a way that allows easy read access by all appropriate staff. At the same time, the information held in a WSP is capable of misuse by unauthorised persons and it is thus important to ensure that correct levels of security, confidentiality and access control for updating are maintained.

It must also be recognised that risk assessment is a dynamic process; the level and type of risk can change over time, as may the water quality goals. Thus the risk assessment must be reviewed and updated on a regular basis and following serious incidents or near misses.

Table 3.1 Examples of risk controls

Hazard	Risk	Control measure	Control point and critical limit	Monitoring	Corrective action
Pathogens from grazing animals around small spring source.	Defined by utility	Fence around catchment area	Integrity of fence.	Weekly by operator.	Immediate repair.
Contamination of treatment chemicals.		Quality control by manufacturer.	Batch contaminant limits agreed with manufacturer.	Each batch analysed by manufacturer and results provided to utility. Basic checks by the utility on delivery.	Batch rejected by manufacturer or utility.
Contaminant ingress during mains repairs.		Defined hygienic working procedures.	Failure to comply with procedures.	Supervisor to inspect repairs on a random basis.	Staff suspended from undertaking specific activity until retrained.
Lead leaching from consumer plumbing.		Maintain pH at consumer taps within defined limits.	Defined pH target ex works.	Continuous monitor on final treated water.	Immediate investigation and rectification.

3.3 ASSET MANAGEMENT

3.3.1 Introduction

The effectiveness of a water utility's operation, including the quality of drinking water, is heavily dependent on a good understanding of its assets and their condition and performance. Asset Management (AM) is a systematic approach to managing service assets in order to minimise costs over the useful life of the assets while maintaining acceptable level of service to customers. Similar to a WSP it is:

- Systematic and structured.
- A cycle of regular review and updating.
- A preventive, risk based, management oriented approach.
- Cost effective.

Both asset management plans and water safety plans are highly complementary. The only difference is one of focus. For example:

- WSPs focus only on drinking water quality, whereas AM addresses other service levels in addition.
- AM deals only with risks associated with water treatment and water distribution systems, whereas WSPs deal with water quality risks from catchment to consumer.

But both rely heavily on a good knowledge of water utility assets. Thus effective implementation of a WSP is also dependent on concurrently implementing effective asset management. The general approach to asset management is summarised in the Figure 3.2 below.

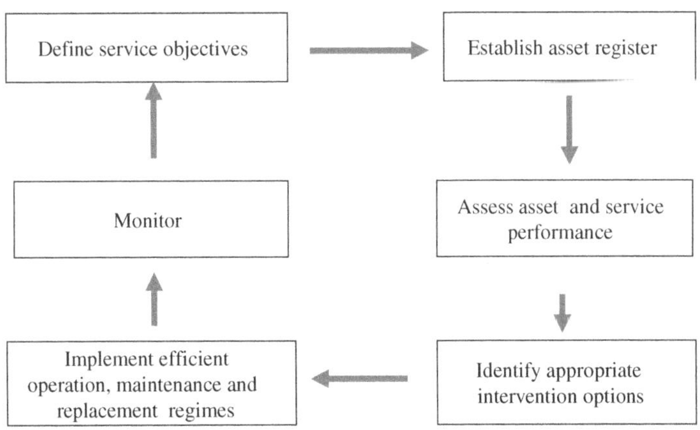

Figure 3.2 The general approach to asset management

A detailed description of asset management is outside the scope of this book but much more information can be found in literature references (Banyard and Bostock 1998, Burn et al 2007, Kleiner and Rajani 1999, IPWEA 2006, Marlow and Burn 2008, Moglia et al 2006, Urquhart 2006a/b). However, some of the key features which relate to WSPs are set out below. Additionally, the importance of asset management is specifically mentioned in Chapters 5 and 6 dealing with treatment and distribution WSPs respectively.

3.3.2 Asset management and water quality

AM covers all service objectives, including water quantity and reliability as well as water quality. The asset service levels for water quality should parallel those for the water safety plan (see Chapter 1.3). As a minimum this should include health protection and compliance with statutory water quality standards, but could also cover for example water acceptability to consumers, measured through complaints or surveys, and the risk of failure of key assets which might impact on water quality, e.g. chemical dosing pumps at water works.

A critical first step in effective asset management is to establish an asset register, which is a structured and detailed inventory of all the utility assets. This should be as complete as possible and capture all available relevant information in a standardised way. Such information could include asset location, type, manufacturer, age, availability of spare parts, maintenance records, and history of associated water quality or service failure.

All the identified assets should then be assessed for both physical condition and performance using standardised criteria. Both are important since it is common experience that some assets might be in good physical condition but still perform badly in an operational sense. Also, age on its own it not necessarily an indicator of performance since some newer assets might in practice to be found to perform worse than older ones. The process and procedures for assessing condition and performance should be as consistent as possible, but must also take into account the experience of front line operators who manage the assets on a day to day basis.

In the same way as for water quality risks under a WSP, assets can be prioritised for intervention using condition and performance in a 5x5 matrix. This allows utilities to identify the assets which perform worst and/or are in the poorest condition in order to cost effectively determine the priorities for replacement, repair or enhanced maintenance.

Having identified the condition and performance of assets, a cost effective long term asset maintenance regime can be established which delivers the required level of consumer service. In water quality terms it is particularly important to determine those critical assets which are important for the continued delivery of good safe water to consumers. In such cases higher levels of preventative maintenance may be employed and/or additional standby capacity introduced to ensure that the overall asset continues to perform effectively even if one component fails.

Finally, as with WSPs, it is important that the asset management plan that has been developed is regularly monitored to check that the identified service levels have been achieved. Also the plan needs to be updated and reviewed at appropriate regular intervals.

3.4 TRAINING AND AWARENESS

Specific training of staff to make sure they have the competencies, skills and knowledge necessary to carry out the tasks required of them is a normal part of any utility's work. As described below, particularly for operational and technical staff this is essential to ensure operational effectiveness and the provision of good safe drinking water quality.

However, as highlighted in chapter 2, specific awareness of the importance of drinking water quality to public health is also an important part of developing a culture among the organisation and workforce that is committed to the principles of risk management and water safety plans. Such awareness training can take a variety of forms and each utility will have to develop its own approach. However it could include:

- Making sure that all staff have a basic understanding of the utility's risk management strategy including the water safety plan.

- Making sure that staff in different parts of the utility understand the importance of drinking water quality and, depending on their role, the part that they personally have to play in public health protection.
- Making sure that all staff that visit operational sites or travel around the drinking water catchment areas are trained to look out for and report back on occurrences which might pose a risk to water quality.
- Ensuring that customer service staff understand the basics of water quality and are trained to respond effectively to consumer complaints (see Chapter 7). It is particularly important that such staff are aware of the need to promptly report any potentially unusual water quality complaints that might provide early indication of a serious problem (see Section 3.6 below).

3.5 OPERATING PROCEDURES

3.5.1 Introduction

Critical to effective water quality management is the implementation of properly structured and documented operating procedures covering all aspects of the water supply system. A detailed description of this is outside the scope of this book but the following sections highlight some particularly important aspects which need to be considered.

3.5.2 Standardised operating procedures

A standard operating procedure (SOP) is a written document or instruction detailing all steps and activities of any business process that could affect the quality of the product or service being provided. The establishment of SOPs is increasingly common practice for water suppliers; the best examples of which are developed in consultation with key stakeholders and consider the entire water supply chain from catchment to consumer. Furthermore, it is important that formal documentation and clear allocation of responsibilities of such procedures is in place.

SOPs are a key part of any water safety plan since they help to ensure that there are clearly defined operating procedures for all staff involved in managing water quality risks from catchment to consumer. The procedures usually cover things such as:

- What tasks need to be carried out?
- At what frequency?
- How the task should be carried out?
- Which people are authorised to carry out the task?
- What level of training/certification is necessary?
- What records need to be kept?
- What to do and who to inform about unusual or unexpected events?

SOPs are usually documented in an operations manual which provides an overview of the utility's operations and creates an important record of the guidelines, procedures, and standards by which water quality and other service standards are to be secured. The operations manual can also serve as a useful training tool for new staff.

It is important that SOPs are regularly reviewed and updated. For example, after a water quality incident or near miss it will be necessary to check the SOPs to see whether improvements need to be made.

As a complement to an operations manual, it is increasingly seen as good practice to also develop a series of operational task schedules which set out the various recurring routine activities that need to be carried out. This can be used to plan the work of operational and other staff and also provide a system to record the fact that the tasks have been carried out at the correct frequency and to a proper standard. Such a system can be particularly useful to ensure that key tasks are not overlooked, for example when staff change shifts, or are temporarily replaced by other staff to cover sickness or holiday.

SOPs should also cover monitoring and data storage. Consistent and structured acquisition and storage of a range of data and information is crucial, not only to good water quality management but all other utility management functions including asset management. The SOP should therefore include such things as what data needs to be captured, at what frequency, to what accuracy, by whom, what data integrity systems are to be used, and how it is stored and subsequently accessed.

3.5.3 Permits to work

There are a number of important operational tasks which usually require staff to be specially trained or qualified before those tasks can be carried out without a significant risk to water quality or other important outcomes such as health and safety, or consumer service standards. In such cases it is common for these tasks to be covered by specific "*permits to work*". This is a requirement that only staff or contractors who are specifically authorised are allowed to carry out such tasks. Use of such "*permits to work*" can provide a useful control component for WSPs. A good example could be the designation of staff and contractors authorised to work on restricted operations (see below). However, it could also apply, for example, to those designated to make certain changes to water treatment controls or operate sensitive valves within the distribution network.

3.5.4 Equipment calibration and testing

Most modern water supply assets have a range of automatic dosing, monitoring and control equipment. Such equipment is often part of the critical control point for ensuring high standards of water quality, e.g. disinfection, and other key water treatment processes. Standard Operating Procedures should include regular maintenance and calibration of such equipment to ensure that it continues to operate as intended and that outputs are reliable. After any significant repair, modifications or upgrade it will be necessary to electronically test the system "*from end to end*" to ensure that the instrument is functioning properly and that the correct signals and alarm conditions, if present, are being transmitted.

3.5.5 Hygienic work practices

The severe public health consequences of water supply contamination, particularly for example ingress of faecal or other harmful material, means that it is particularly important to have in place specific operating procedures to avoid such risks. These can be formalised and documented under the general term Hygiene Code and should be included as part of the training programme for all relevant employees and contractors. The Hygiene Code should include:

3.5.5.1 Definition of restricted areas/operations

Water which has been fully treated and put into supply is usually in sealed pipes/tanks and mostly under pressure. It is thus less likely to suffer contamination, provided the integrity and pressure of the system is maintained. However, there are certain parts of the system where there is a much higher risk of contamination and where special precautions will be necessary. These will vary according to local circumstances, but would typically include all areas where there is the potential for direct contact with untreated groundwater sources, with partially or fully treated water in waterworks, or within the distribution network. Each supply system should have such areas clearly defined, and access controlled through locks and signs. Any work carried out in restricted areas should be classified as restricted operations.

3.5.5.2 Designation of personnel who can work in restricted areas

Only specially trained staff should be authorised to work in restricted areas/operations. Normally after such training the individuals concerned would be issued with a special permit which confirms that they have been cleared to work in such areas. They should also be reminded of their obligations to notify their employers in situations when they might present an increased risk to the water supply. This would normally be associated with suffering from diarrhoeal illness or living in close proximity to known carriers of certain high risk water-borne diseases as described below. The training and permit should be updated at appropriate intervals. In addition to water utility staff, any contractors who might be required to work in restricted areas or operations must also be covered by the same training and permitting procedures.

3.5.5.3 Medical screening

Although all work in restricted areas must be subject to strict hygiene procedures, any staff who contract disease involving diarrhoea or fevers, or who might be asymptomatic carriers of such disease must report this to their managers and not be allowed to work in restricted areas until confirmed to do so by medical practitioners.

3.5.5.4 Hygienic working procedures in restricted areas

All work in restricted areas should be subject to specific precautions to avoid water contamination. This should include: personal hygiene; strict rules on defecation and urination; cleanliness and disinfection of clothing, tools, and equipment; use of chemicals; notification of any chemical spill or contamination that does inadvertently occur; and disinfection requirements after the work is complete. Specific examples of hygienic work practice are given in later chapters.

3.6 EMERGENCY PROCEDURES

3.6.1 Introduction

However well managed, any water supplier will have to deal with incidents of varying levels of severity potentially impacting service to consumers. The trust that consumers have in their water supplier, and sometimes the overall reputation of the supplier, depends critically on the way that they respond to such incidents, particularly those which threaten water quality. The resources available to manage incidents will vary considerably from utility to utility. Many of the principles are similar but each utility will need to develop an approach which is appropriate to its needs. This section sets out a framework for managing incidents which can be adapted to suit local needs and circumstances. More information is given in references (Gray and Thompson 2004 and 2009, Thompson and Gray 2006).

In a worst case situation, incidents can typically go through three levels of severity as summarised in Figure 3.3. The time and resources needed to return to a steady state depends on the ultimate scale and severity of impact. The aim of incident response procedures should be to deal with the vast majority of incidents as quickly and effectively as possible in order to minimise the impact on consumers. For those that do impact on consumers or the wider community, the aim should be rapid recovery to normal levels of service, coupled with effective consumer communication, so that the response is seen by all stakeholders as professional and responsive. Incidents which are well managed can often enhance the reputation of the supplier.

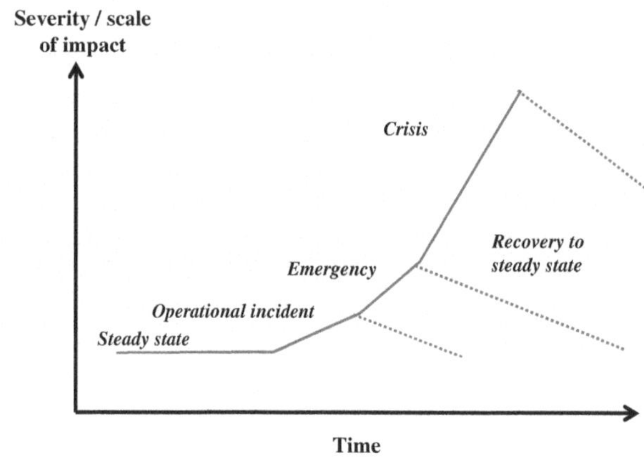

Figure 3.3 Typical stages of an incident

3.6.2 A scalable incident response system

Because of the wide variety of incidents that can occur, the response system has to be well structured but flexible. This requires:

- Setting up an incident management structure appropriate to the size of organisation which operates 24 hours a day/7 days a week.
- Ensuring that all staff are aware of the importance of identifying and notifying potential incidents as soon as possible.
- Procedures for early assessment by experienced managers/operators to categorise the level of incident and determine the appropriate initial response.
- Depending on the type of incident and potential scale of impact, procedures for escalation of responsibility to more senior staff.
- Clearly defined responsibility for initiation of formal emergency status including, for example, bringing in additional resources.

A typical process for this is summarised in Figure 3.4 below:

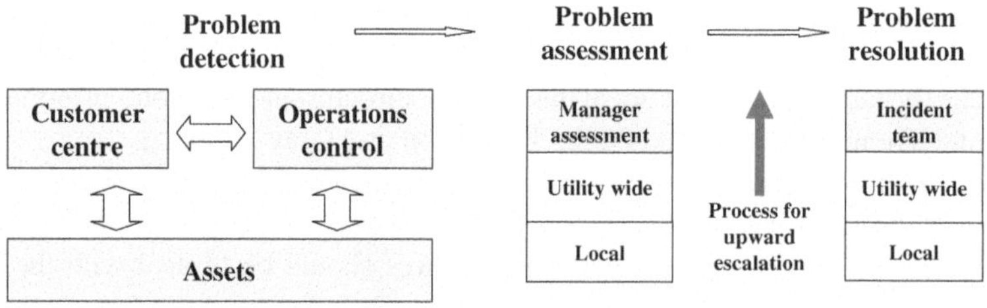

Figure 3.4 Outline of a typical incident management process

3.6.3 Managing the early stages of an incident

Many serious incidents start as small ones which escalate out of control. It is therefore essential to have early detection and rapid response to problems in order that the situation can be stabilised and recovered as quickly as possible. Information can come from a variety of sources including:

- Internal monitoring systems.
- Consumers/public.
- Other external organisations (e.g. catchment managers).

Thus it is essential to have procedures which define different types of event or incident, and the appropriate response, as well as clearly defined internal/external communication processes. However, procedures should be based on the assumption that information will often be limited or inconsistent, particularly in the early stages of an incident.

3.6.4 Involvement of external organisations

During a serious incident there are likely to be a range of other organisations involved including, for example, health and municipal authorities, catchment managers and regulators. It is important that these organisations are identified before an incident occurs and good working relationships developed with them. A simple checklist of these may be useful as a reminder within the procedure. Water suppliers should work with them on a range of issues where there are shared interests and regular contact should be maintained. Relevant external agencies should be involved in the preparation of the incident management procedures, including how to get in touch with them 24/7 if problems occur.

3.6.5 Training and rehearsal

There is nothing like the real thing to test procedures and staff competence. Therefore, while it is time consuming and resource demanding, it is considered essential to run regular exercises to test incident management procedures. These can either be based on desk top or real-time scenarios and wherever possible should involve relevant external agencies. At the end of each exercise there should be a thorough review, ideally using people not directly involved in the management of the particular incident, to identify areas of weakness where improvements might be required.

3.6.6 Dealing with the media

It is common for newspapers, radio or TV to cover significant water supply incidents. Such coverage can be a key part of consumer communication but depending on how this is handled can either enhance or damage the reputation of the water supplier in managing the incident. Thus, incident management procedures must include prior identification of people who can speak for the water utility, who are properly trained in working with the media and available 24/7. It also requires internal procedures for agreeing media messages which are:

- Factual, realistic, appropriate and timely.
- Agreed with external parties where necessary.
- Based on best currently available information.

3.6.7 Incident management plans and their documentation

Effective incident response requires good advance planning, properly trained staff and well documented, readily accessible procedures. However, whilst management plans need to be comprehensive, they also need to be flexible enough to cope with the unexpected.

Documenting of incident management procedures and related information is vital, but this must be kept up to date and regularly reviewed. Information must always be available to all relevant staff on a 24/7 basis. In larger organisations this means that such information is best stored electronically, although in smaller utilities paper records can also work well, provided they are up to date. The response to routine incidents is normally covered in Standard Operating Procedures (SOPs), but many water suppliers have procedures for more serious incidents included in an incident manual, which is often broken down into sections covering:

- Emergencies.
- Crises.

A check list of typical items needed for an emergency management plan is given below. It must, though, be recognised that such information is capable of misuse by unauthorised persons. Thus it is important to ensure that correct levels of security and confidentiality are maintained.

3.6.7.1 Roles and responsibilities

The plan needs to clearly define the roles and responsibilities of each person involved in the management of an emergency, including an incident manager plus a range of other key support roles. The responsibility and procedure for initiating the establishment of a formal emergency team also needs to be documented.

3.6.7.2 Process for mobilising resources

An emergency usually requires mobilisation of a wide range of additional resources, with the first few hours of the response being particularly important. The plan should thus define the method and timing of the mobilisation process. This should take into account the fact that, during normal working hours, teams can be in place fairly quickly, but at night, weekends or during holidays it will take a lot longer to contact people and for them to make the journey to where they are required. Thus, up to date call-out arrangements and out of hours contact lists for key staff are important.

3.6.7.3 Support and back up for all key staff

All staff involved in emergency management should be trained in their specific duties. This should include participation in exercises to test emergency procedures. If it is expected that an incident will last for more than 6–12 hours then planned staff rota and rest systems need to be established, including maximum shift periods for effective working. As well as those staff involved in the emergency, it is important to ensure that there are also plans to maintain adequate staffing for vital ongoing routine operational activities.

3.6.7.4 Emergency management centre

The emergency management team should, ideally, be established in a room which is specially identified and equipped for this purpose. This could include dedicated phones and other communication equipment, adequate working space and appropriate administrative support. Consideration also needs to be given to the potential need for other logistical support on a 24 hour basis including refreshment for team members, security and access to buildings and other practical issues. When not in use, the room can be used for other purposes, but it has to be understood that when an emergency incident is declared, the room and all the facilities are to be dedicated to the Incident Management Team immediately.

3.6.7.5 Access to records and plans

The emergency management team will need to have easy access to relevant maps, records, plans, operating procedures and other relevant information. In particular, information should be available on all the major assets, including layout drawings of the treatment works and pumping stations and plans of the trunk main pipelines and the distribution system. These may be held electronically or on paper. Similar information may need to be available at other operational centres dealing with the emergency. It may be necessary for other more detailed information to be accessed during the emergency, but if it starts during the night or at weekends the response time may be slower.

3.6.7.6 Access to emergency equipment and spares

Depending on the cause of the problem, it is likely that rapid access to a range of specialist emergency equipment and critical spare parts will be needed. Lists of such equipment should be drawn up, indicating whether it is specifically held for emergencies only, used operationally, or borrowed or hired from other organisations. For each piece of equipment, details should be recorded of key facts such as normal location, responsible officer, mobilisation time, method of transportation, training required by driver/operator, fuels and other items required to operate the equipment, details of all pipe and cable connections.

3.6.7.7 Emergencies impacting public health

For any emergency impacting, or potentially impacting, public health, even faster and more effective response will be required. During such incidents close and early liaison needs to be maintained with public health authorities. It is helpful to agree in advance the criteria for this to occur. Appropriate arrangements will need to be made to access specialist medical advice on such matters as public health impact and appropriate responses to both microbiological and chemical contamination.

Necessary measures to protect public heath will vary depending on the situation. As far as possible emergency plans should include procedures for those responses which are most likely, including, for example, notifying consumers of advice to boil water before use, or temporary distribution of water via tankers or other means. Before issuing any "*boil water*" or similar advice to consumer, the criteria for letting consumers return to normal use practice needs to be agreed with the health authorities.

It must also be remembered that during emergencies affecting public health, additional monitoring programmes will need to put in place and mechanisms established for rapid analysis and dissemination of the results to all interested parties.

3.6.7.8 Communication

Effective internal and external communication procedures are a vital part of an emergency plan. A list of all organisations, officers and individuals who may have to be contacted during an incident should be kept up to date, and made available in the Incident Management Room. External communications will typically be required with municipal and regional authorities, regulatory authorities, health authorities, other utilities (e.g. power suppliers), the emergency services and the media.

During a major incident regular and effective communication with consumers is of vital importance. This might be proactive via TV or radio to inform the public of the ongoing situation and advise them of what they should do, and reactive to deal with consumer enquiries. It may be helpful to develop in advance typical questions and answers that might be required for such consumer and media communication.

During such an incident the number of telephone calls or other contact from consumers will increase substantially. It is therefore important to organise an emergency consumer response system which has the ability to receive, handle and process the high number of calls that might be expected. The system design should allow for the transition from a normal to the emergency situation. Frequent briefing should be given to all the team members to ensure they are kept up to date on the changing situation.

3.6.7.9 Incident logs

It is of vital importance to maintain an incident log which records all key decisions taken, contacts made, information received and instructions issued and the time at which these occurred. Such information will be invaluable during any incident debrief.

3.6.7.10 Stand down procedures

Once an emergency has been dealt with it will be necessary to formally notify all the people and organisations involved directly in the incident, the press and consumers that it has ended. Where necessary, appropriate instructions should be given to staff on all the actions still to be taken and the timescales for this. There should be a clear process for a smooth transition from emergency management situation back to the organisation's normal management structures and procedures. Once an incident has ended, a full and detailed review of any incident should be carried immediately to identify any lessons learned and things that could be improved. Many organisations use an independent person to carry out the review and make the results available to their consumers to demonstrate open management.

3.7 CONTRACTORS AND SUPPLIERS

It is important that WSPs cover not only staff, but also all relevant contractors and suppliers. A brief summary of the general approach is given below and more detail is provided in other relevant chapters.

3.7.1 Contractors

It is common for utilities to use a range of contractors to carry out a range of tasks, many of which could impact on water quality. It is thus important that as part of the contractual agreement, staff employed by a contractor are required to have the same level of training, and operate to the same or equivalent procedures, as those used directly by the utility. Controls included in the WSP to manage identified water quality risks will also need to cover contractors.

3.7.2 Suppliers

Water utilities will have a large number of suppliers who provide a wide range of goods and services. Where those might potentially impact on water quality they need to be identified as part of the water safety plan and appropriate controls included. These are covered in more detail in later Chapters, but for example might include suppliers of pipes, fittings and other material used in direct contact with treated water, as well as water treatment chemicals. Assessment of risks to water quality need to include, not only the quality and suitability of materials and services supplied, but also how to ensure the reliable availability of critical supplies such as treatment chemicals, for example, if the main supplier has a fire or other incident at their premises.

3.8 STAKEHOLDER PARTNERSHIPS

The Bonn Charter emphasises that whilst water suppliers are central to the provision of good safe drinking water, this has to take place within a broader legal, institutional, and financial framework which is the responsibility of Government and many others. It reminds us that an integrated catchment to consumer approach to water quality risk management will require close co-operation and partnership between Governments, water suppliers, health agencies, environmental agencies, land users, contractors, plumbers and consumers themselves. The institutional arrangements and type and nature of the organisations involved will vary considerably between countries and regions, but it is important that, as part of the development and implementation of a WSP, water suppliers work with many other partners to ensure that all are committed to the same goal of good safe drinking water.

It takes time and effort to develop such relationships and so the utility will need to identify the priority that it needs to assign to different stakeholders. In some cases, particularly when dealing with national stakeholders, it may be better for a number of utilities to work together as part of a regional or national water supplier association. Figure 3.5 below summarises the typical stakeholders with which water suppliers may have to develop partnerships. More details of how to work with individual stakeholders are given in relevant chapters.

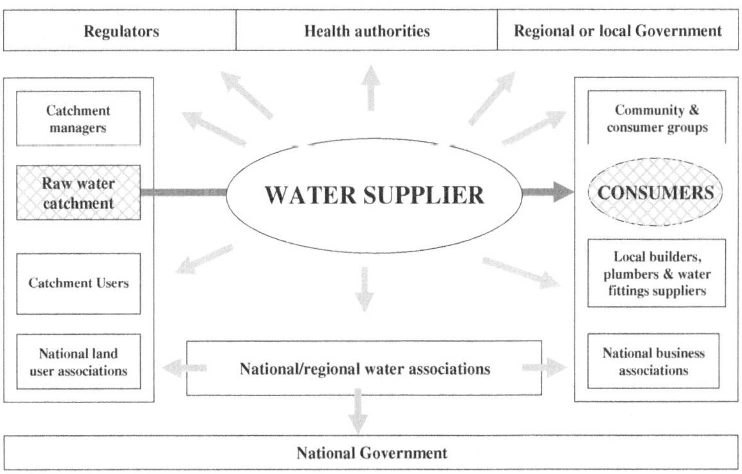

Figure 3.5 Stakeholders with which water suppliers may have to work as part of their WSP

3.9 RELIABILITY OF SUPPORT INFRASTRUCTURE

3.9.1 Introduction

There are a number of risks that can impact across all aspects of a water utility's operations, including water quality. Assessing and mitigating such risks may already be covered by the utility's overall risk management strategy, but they also need to be linked specifically to the development of a water safety plan. This section briefly summarises some of the risks which need to be considered. In some cases these are also considered in more detail in other chapters. Many of the issues raised will almost certainly also need to be covered in the water supplier's emergency management procedures (see earlier Section 3.6).

3.9.2 Power and energy supplies

Electrical power and other energy supplies are a vital part of all water utility operations. The reliability of power supplies will vary considerably in different countries. In developed countries supplies are usually fairly reliable, but can be impacted by adverse weather or other events, particularly in remote rural areas. In less developed countries power supplies can sometime be more erratic. Water quality risks created through loss of power supplies will need to be assessed as part of any water safety plan and appropriate mitigation measures included in overall risk management strategies. The most cost effective risk mitigation measures will depend very much on local circumstances but could, for example, involve making arrangements with energy suppliers for priority to be given to water supplies as part of emergency response, provision of reliable standby power generation capability, or in particularly high risk situations providing separate dual power supplies to critical operational sites. Surge protection and uninterrupted power supply (UPS) back up of certain sensitive and critical instrumentation and control equipment may also be prudent.

3.9.3 Telemetry and remote sensors

Many aspects of modern water quality management rely on remote sensors and telemetry of critical data and information from unmanned sites. The reliability and impact of losing these telemetry systems or the generation of false readings or alarm situations will need to be assessed as part of any water safety plan. For example as well as physical damage, such systems can often be severely impacted by bad weather, including electrical storms.

3.9.4 Flooding and other extreme weather events

There are a wide range of extreme weather conditions that can create risks to water supplier's operations including drinking water quality. These will need to be taken into account as part of the development of the water safety plan as well as any broader risk management strategies. For example, heavy rainfall can rapidly degrade raw water quality and threaten water treatment capacity. In extreme cases flooding can lead to loss of operational resilience with some sites having to close or operate at reduced capacity. In drought conditions availability of raw water supplies will be at risk, as may be the quality of the raw water.

3.9.5 Site security

Ensuring site security against vandalism or malicious intruders is an increasingly important part of any water utility's broader risk management strategy and should also be considered as part of the development of a water safety plan. The degree of security necessary can only be assessed locally, taking into account a range of factors and the sensitivity of each site, both operationally, and in terms of vulnerability to unauthorised access.

3.9.6 IT availability and back up

The operation of most modern water utilities is underpinned by a complex range of IT systems. Planning to cope with the failure of any IT system would normally cover such issues as physical security of IT equipment and data, secure back up of systems and data, and disaster recovery procedures which allow systems to be restored as quickly as possible after loss or damage, for example, through external 3rd party hosting. Water quality risks created through loss of systems or data will need to be assessed as part of any water safety plan and appropriate mitigation measures included in data security strategies.

3.9.7 Laboratory facilities

Microbiological and chemical laboratory facilities are an important part of a water supplier's operation. These may be operated by the utility or in some situations, some or all, of the analysis may be subcontracted to another laboratory. WSPs will need to take into account the impact of loss of such laboratory facilities, for example after a fire or other occurrence. Plans will also need to be in place to ensure access to additional laboratory facilities during periods of high sampling throughput and emergency turnaround times, for example during water quality emergencies.

3.10 MONITORING, DOCUMENTATION AND REPORTING

3.10.1 Introduction

Effective up to date monitoring, documentation and reporting underpin good utility management generally, and in the context of water safety plans, water quality management particularly. Thus it is important to ensure that system knowledge is captured, managed and made easily accessible to all relevant staff. However, this is often an area which causes problems for water utilities. If it is too detailed then it may hinder progress by becoming too resource intensive and bureaucratic. If it is too simple then there is a risk of it being insufficiently useful and missing key areas of weakness.

The approaches that can be used will be specific to the needs of the individual utility and are outside the scope of this book, but the following sections highlight some of the key aspects that need to be considered.

3.10.2 Documentation

Documentation is a key part of any modern water utility. In respect of water quality management, as well as that necessary for the water safety plan itself, a wide range of other forms of documentation will be necessary and need to be cross referenced to the WSP. This could cover such things as:

- Asset registers which record the location, condition and performance of each asset.
- Standard operating procedures for a wide range of tasks across the utility.
- Laboratory analytical quality control procedures.
- Emergency management procedures.
- Water quality data, both for regulatory and operational purposes.
- Data on incidents and near misses potentially impacting water quality.
- Plant operating data, including for example on site monitoring data, alarms, chemical usage.
- Staff training, competency and certification records.
- Contractual and purchasing records.
- Customer complaint and other data.

Effective storage and access to such data is critical. In most utilities the volumes of data involved will be substantial and thus will require very effective information management, document control procedures and IT systems. It will also need to ensure that there is appropriate cross links between the datasets. Providing easy access to the correct data by all relevant staff at all times, whilst balancing the need for data security, provides additional challenges.

3.10.3 Water quality sampling and analysis

As described earlier, water quality sampling and analysis is a vital part of any water supplier's operation, whether this is carried out within the utility or subcontracted to other service providers. As well as carrying out routine verification tests to ensure that water quality is meeting statutory and other water quality requirements, laboratory analysis is important for a range of other operational purposes including during water quality emergencies.

It is important to ensure that laboratories operate to high standards of best analytical practice, including effective quality control procedures both for sampling and analysis.

3.10.4 Monitoring and reporting

Monitoring and reporting is a key part of water quality management and water safety plans. However, to be effective this requires much more than just providing reports on laboratory monitoring data against regulatory parameters (verification monitoring). It also needs to support a comprehensive risk based approach and be geared to the need of the intended audience both internally and externally.

The data collected and reported for water quality management purposes could fall into a number of categories as set out below:

- Compliance (verification) reporting - results of water quality monitoring required by legislation and commentary on any failures and necessary action taken.
- Water quality incidents and near misses. This could include both incident trends and reviews of major incidents to identify lessons learned.
- Consumer acceptability measured through such things as complaints, surveys or other means.
- Water safety plan management and updating.
- Effectiveness of critical control points - significant deviations or adverse trends in operational monitoring and action taken as a result.
- Progress with water quality improvement plans.
- Results of internal and external water quality audits.
- Laboratory performance and information.
- Asset management data.
- Potential future changes to regulation which might require additional investment or planning.

Internal reporting procedures should be open and transparent and be carried out with full involvement of all staff at all levels. The identification of problems should be encouraged with an honest and open approach. As described in Chapter 2, the development of an organisational culture which supports such a risk management approach will greatly improve the effectiveness of overall water quality management. Reporting will need to take place at a variety of levels from local operating teams to senior managers and boards. This will require a well designed and efficient data management system which provides the appropriate level of information for the audience. At senior manager level, data normally needs to be provided in summary format and many utilities use a "*traffic light*" reporting approach which colour codes key data. Where necessary such data can be designed so that data showing amber or red can be expanded to provide more details. For example:

- Green: results OK.
- Amber: adverse trend.
- Red: needs urgent attention.

Chapter 7 deals in more detail with reporting to consumers and other external groups.

REFERENCES

Bartram, J., Corrales, L., Davison, A., Deere, D., Drury, D., Gordon, B., Howard, G., Rinehold, A. and Stevens, M. (2009) Water Safety Plan Manual: Step-by-step risk management for drinking-water suppliers, World Health Organisation, Geneva.

Banyard, J. K. and Bostock, J. W (1998) Asset management - investment planning for utilities, *Proceedings of institution of civil engineers, Civil engineering*, **126**, 65–72.

Burn, S., Marlow, D., Moglia, M. and Buckland, P. (2007) Asset management for water infrastructure, *Water asset management international*, **3**(1), 12.

Davison, A., Howard, G., Stevens, M., Callan, P., Fewtrell, L., Deere, D., Bartram, J., (2005) Water Safety Plans: Managing drinking-water quality from catchment to consumer, World Health Organization, Geneva.

Godfrey, S. and Howard, G. (2004) Water Safety Plans (WSP) for Urban Piped Water Supplies in Developing Countries, Water, Engineering and Development Centre, Loughborough University, UK.

Gray G. and Thompson K. C., (2004) Water Contamination Emergencies - Can we cope?, Royal Society of Chemistry, London.

Gray G. and Thompson K. C., (2009) Water Contamination Emergencies - Collective Responsibility, Royal Society of Chemistry, London.

Hokstad, P., Røstum, J., Sklet, S., Rosén, L., Pettersson T.J.R., Linde, A., Sturm, S., Beuken, R., Kirchner, D., Niewersch C. (2009) Methods for risk analysis of drinking water systems from source to tap - Guidance report on Risk Analysis, Techneau Report D.4.2.4 http://www.techneau.org/fileadmin/files/Publications/Publications/Deliverables/D4.2.4.pdf (accessed 2 December 2010).

IPWEA - Institute of Public Works Engineering Australia (2006) International Infrastructure Management Manual.

Kleiner, Y. and Rajani, B. (1999) Using limited data to assess future needs, *American Water Works Association Journal*, **91**(7), 47–61.

MacGillivray, B.H., Hamilton, P.D., Strutt, J.E. and Pollard, S.J.T. (2006) Risk analysis strategies in the water utility sector: an inventory of applications for better and more credible decision making, *Crit. Rev. Environ. Sci. Technol.* **36**, 85–139.

Marlow, D. and Burn, S. (2008) Effective use of condition assessment with asset management, *American water works association. Journal*, **100** (1), 54.

Moglia, M., Burn, S. and Meddings, S. (2006) Decision support system for water pipeline renewal prioritisation, *Electronic Journal of Information Technology in Construction*, **11**, 237–256.

Rosén, L., Hokstad, P., Lindhe, A., Sklet S., Røstum, J. (2007) Generic Framework and Methods for Integrated Risk Management in Water Safety Plans, Techneau report D4.1.3 http://www.techneau.org/fileadmin/files/Publications/Publications/Deliverables/D4.1.3.pdf (accessed 2 Dec 2010).

Thompson K. C. and Gray G., (2006) Water Contamination Emergencies - Enhancing our response, Royal Society of Chemistry, London.

Urquhart, T. (2006a) Asset management around the world: the expertise of Australia/New Zealand, the UK and the US, *Water asset management international*, **2**(1), 3.

Urquhart, T. (2006b) Incorporating condition assessment into a comprehensive asset management programme, *Water environment foundation: WEFTEC*, Oct. 2006, Dallas TX, Water Environment Foundation, 4198.

WHO (2011) Guidelines for Drinking Water Quality, 4th Edition, World Health Organisation, Geneva.

Chapter 4

Developing a catchment water safety plan

José Vieira, Bob Breach and Ricardo Hirata

4.1 FACTORS UNDERPINNING THE DEVELOPMENT OF CATCHMENT WATER SAFETY PLANS

4.1.1 Introduction

The quality of raw water is a key element when selecting a source for any drinking water supply system. Water use, land use and polluting human activity in the catchment area all have significant impacts on surface and groundwater quality, and thus the level and complexity of treatment plant necessary to ensure that the water leaving the works is safe and acceptable to consumers.

Understanding catchment characteristics and/or activities potentially impacting on raw water quality and availability is thus of paramount importance to ensuring drinking water safety. Protection of raw water sources should be seen as the first, and often the most important, barrier to prevent microbial, chemical and radiological contamination of drinking water.

The development of a catchment WSP based on risk assessment and management procedures, is very dependent on the quality and quantity of relevant information available. Particular effort is required to collect information on catchment characteristics (e.g. geology, hydrology, meteorology, land use, competing water uses), surface water bodies (e.g. flow rate, water quality and seasonality) and groundwater (e.g. aquifer flow rate, flow direction and aquifer vulnerability to pollution).

This chapter sets out some of the factors underpinning catchment management, provides advice on developing a catchment WSP in conjunction with external stakeholders, describes methodologies for assessing and managing both surface and groundwater catchments and briefly summarises common catchment risks and how they might be mitigated.

4.1.2 Characteristics of different catchment types

The characteristics of different types of catchment vary considerably, including the response to different polluting activities or events, the time taken for pollution to reach the water abstraction point and the ease and timescale with which pollution prevention can be realistically carried out. Additionally, the pathways by which pollutants move from the pollution source to the abstraction point can be complex and need to be fully understood before the risks can be properly assessed and effective mitigation options pursued. All of these factors and others need to be taken into account when developing a catchment WSP. The phrase "*know your catchment*" is a vital one for any water supplier.

The majority of catchments can typically be divided into three main types, but within each type there can still be considerable variation in characteristics and behaviour:

- *Groundwater*
  The passage of water through the soil layer and underlying rock strata to the water table can attenuate some, but not all, types of pollution and thus treatment for groundwater tends to be simpler than for surface water. Deep groundwater typically has a much slower response to polluting activities on the surface, often many years (Howden et al. 2011). For some shallow or fissured groundwater sources, however, the response can be much faster, sometimes as little as a few days.
- *Upland Reservoirs*
  In some countries water is captured from upland (mountainous) areas and held in reservoirs before treatment. Typically, the less developed nature of these catchments means that the raw water may be less prone to man made pollution than other surface water sources, but still requires more treatment than groundwater. However, livestock grazing and poor moorland management can present particular pollution risks, and if pollution occurs then the reservoir can remain contaminated for long periods because residence times can be lengthy and some contaminants are not rapidly broken down.
- *Lowland surface sources*
  This can include rivers or canals either with or without raw water storage. Raw (untreated) surface water, particularly from rivers, is normally of lower quality than groundwater due to naturally occurring pollutants (see Section 4.5.2), although this can often be made worse by the impact of human activity. Additionally, there are a wide range of anthropogenic (man-made) pollutants that can arise from a number of sources and activities, and which if not controlled could require additional treatment specifically for their removal at waterworks. For this reason, surface water normally requires higher levels of treatment to meet drinking quality standards, compared to groundwater. Typically, this would include coagulation, filtration and disinfection, sometimes coupled with additional treatment by activated carbon or other more specialised processes.

4.1.3 Objectives for catchment WSPs

There are a number of inter-related objectives for developing catchment WSPs (Figure 4.1). All rely on a good knowledge of the catchment.

- *Implementing measures to reduce catchment water quality risks*
  The primary aim of a catchment WSP is to reduce risks within the catchment to protect water quality at the point of abstraction. However, even where effective arrangements for catchment management and control have been implemented, unexpected deterioration in raw water quality can pose a risk to treated water quality. Thus, pollution early warning systems must be developed in all catchments to allow water suppliers to be notified as soon as possible about any pollution events so that emergency procedures can be instituted. This is even more important in the case of groundwater

catchments with no, or only limited, treatment. In such situations early warning systems, based on periodic sample analysis and inspection of the well network, are essential.

- *Monitoring raw water quality and its variability*
  Design of raw water monitoring programmes is helped considerably by having a detailed knowledge of the catchment. This ensures that sampling and analysis, which can be very expensive, are cost effectively targeted at those parameters most likely to be of significance (see later Section 4.2.7).
- *Designing effective treatment barriers*
  Knowledge of the catchment coupled with routine risk-based monitoring programmes, supported by modelling of pollutant sources and transport processes, can help to build up a picture of expected raw water quality and the extent of any quality variability. This allows the treatment plant to be designed and operated with sufficient spare capacity to deal adequately with the poorest predicted raw water quality (see Chapter 5). Groundwater is commonly pumped into water supply systems with no treatment other than simple disinfection with chlorine. In this situation, catchment risk monitoring, assessment and management becomes even more important.

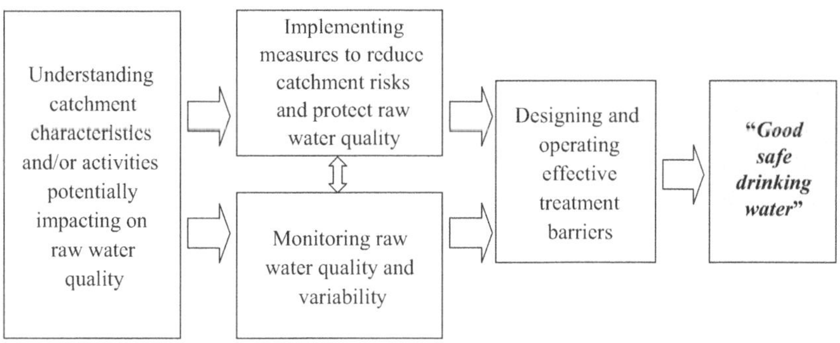

Figure 4.1 Objectives of catchment WSPs

4.1.4 Impact of raw water pollution on drinking water treatment

Knowledge of land use and management, or other activities which might pose a risk to raw water quality, allows catchment managers, in conjunction with water suppliers, to introduce pollution prevention measures to reduce such risks. Risks can impact water suppliers in a number of different ways:

- Where installed treatment can cope with the normal variation in raw water quality, then catchment protection can avoid the likelihood of unexpected pollution events posing a risk to treated water quality.
- Where water quality has already deteriorated, or is deteriorating, to the extent that additional treatment is or may be necessary in the future, then catchment protection may reduce or eliminate the need for new treatment. In this context it is important to recognise that groundwater response time to contaminant source reduction or elimination is much longer than that observed for surface water (Howden et al. 2011).

- Where a works has been abandoned completely because of poor raw water quality, improved catchment protection may allow the source to be cost effectively brought back into use. This might be of particular importance in areas of severe water resource shortage.

4.1.5 Categories of pollution risk

In any catchment, there are a wide variety of potential pollutants and mitigation options. More information is given later in Section 4.5. Their origin can typically be categorised into three main types:

- *Point sources*
  This is where the pollution derives from a single point of discharge such as from a sewage works, factory waste outlet or solid waste disposal facility. This can be mitigated through suitable risk management techniques to avoid leaks or spillages, and appropriate wastewater treatment. Depending on local legislation, this can be enforced through pollution prevention requirements (e.g. bunding of tanks) and use of regulatory "*permits to discharge*" for example, specifying maximum pollution loads. In the case of groundwater catchments, pollution prevention may require on site remediation or hydraulic containment of the aquifer contamination plume, as well as the removal or reduction of the contaminating activity.
- *Diffuse sources*
  This is where the pollution arises from a multiplicity of small sources such as runoff from agricultural land, roads or from domestic waste water e.g. septic tank discharges in areas without mains sewerage cover (Arnscheidt et al. 2007). Commonly occurring diffuse source contaminants include nutrients (especially nitrogen and phosphorus), pesticides and faecal pathogens e.g. from livestock (Kay et al. 2008). Diffuse sources can be much harder to control since, by definition, many different land users may be required to implement significant change in practice.
- *Naturally occurring sources*
  Natural contamination can occur due to local geology and/or soil types and is associated with chemical dissolution of minerals and organic matter. In addition, faecal pathogens may be transferred from land to water, for example from wildlife (McDonald et al. 2008). In some catchments, natural contamination may be made worse by atmospheric inputs, for example the enhanced release of aluminium from acidified soils or the enhanced release of dissolved organic matter from peat soils after severe summer drought (Mitchell and McDonald 1992).

4.2 DEVELOPING A CATCHMENT WATER SAFETY PLAN

4.2.1 A framework for catchment WSPs

Many water supply catchments are complex and water suppliers do not normally have the responsibility, the powers, and often the expertise to directly control activities within the catchment. Development of a catchment WSP is, therefore, usually much more difficult than that for treatment or distribution network WSPs. This section summarises an approach to developing a catchment WSP based on a modified version of the classic WSP cycle of activities (Bartram et al. 2009, Vieira & Morais 2005).

Most, if not all, of the catchment-based activities should be incorporated in a catchment management plan which is developed and jointly owned with all other key stakeholders in the catchment. Increasingly in many countries the concept of "*ecosystem services*" is being used to manage catchments for which the provision of clean water for potable water abstraction is seen as a key element. The plan should encompass a range of activities which are kept under regular review and include the essential elements of risk assessment and risk management as shown in Figure 4.2.

However, experience shows that the larger and more complex the catchment, the more difficult it is to ensure completely effective catchment protection. Water suppliers will thus need to assess realistically the likelihood that catchment protection measures will be fully effective and the necessary timescale. Although catchment protection is by far the best option, if this is not possible within a reasonable timescale, then additional treatment or other measures may need to be considered in order to ensure the quality of treated water.

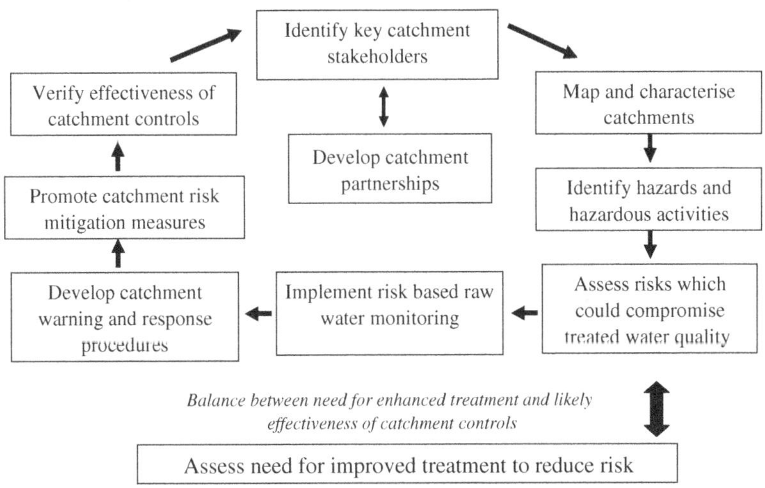

Figure 4.2 A framework for preparing a catchment WSP

4.2.2 Identifying key catchment stakeholders

Although water suppliers do not normally have powers or the responsibility to control activities within the catchment directly, it is essential that they work in conjunction with relevant government agencies, local catchment managers and land users to identify ways that pollution risks can be most cost-effectively managed.

There are a wide range of catchment stakeholders with whom water suppliers may need to develop partnerships to protect water abstractions. They can be broadly divided into (1) regulatory authorities and catchment managers and (2) land users carrying out activities which could pose a pollution risk. The institutional arrangements and legislative framework for managing catchments will vary considerably between countries and this will impact on the way in which such partnerships can be developed. It also means that partnerships often need to be developed at national/regional and local catchment levels (see Figure 4.3).

4.2.2.1 National/regional level partnerships

Since pollution control policy and legislation are usually developed and promoted at national or regional level, water suppliers will need to work together through their national or regional water supply associations to make sure that the interests of water supply protection are properly taken into account. The requirements of water supply protection may be different to that for protection of aquatic biodiversity and this often needs to be explained to policy makers.

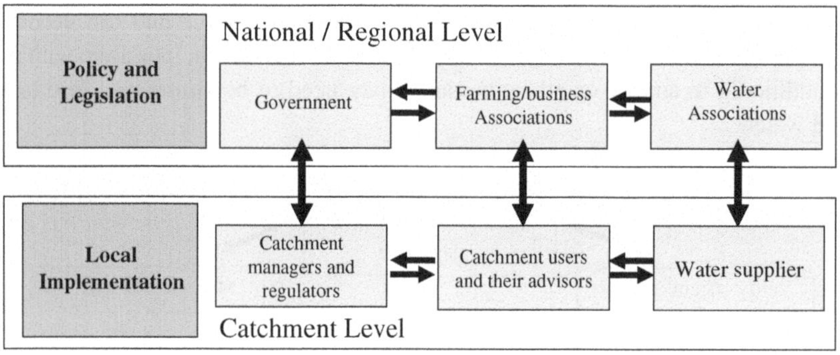

Figure 4.3 Catchment partnerships at different levels

As well as policy partnerships with Government and regulatory bodies, catchment protection can also be enhanced if water supply associations work directly with national or regional farming or business trade associations to jointly develop codes of best practice or certification systems. Local catchment users are much more likely to adopt such measures if they have been developed in close cooperation with their representative professional associations.

4.2.2.2 Local catchment partnerships

Effective local partnerships between water suppliers and land users/catchment managers normally benefit from being developed as part of a broader national/regional policy and legislative framework, although successful local partnerships can also be developed on their own (Stollard and Rickard 2005). Within this framework the purpose of local partnerships is to engage, educate and persuade catchment users to adopt pollution control and mitigation measures specifically designed to protect water abstractions.

4.2.3 Developing catchment partnerships

The approach to working with stakeholders can vary depending on whether they are more likely to give rise to point or diffuse sources of pollution risk. For example, point source discharges from industrial premises or sewage works are often covered by statutory discharge controls which limit the type, quality and quantity of effluent that can be discharged. Where such legislation is properly enforced, any additional pollution mitigation measures may need to focus primarily on avoiding accidental pollution (e.g. as a result of spillage). Where the risk mainly relates to diffuse pollution then many more stakeholders need to be involved and this can increase the difficulty of developing effective partnerships.

It is also important to make a realistic assessment of the cost and complexity of the mitigation measures that are being promoted, since this will determine the approach which needs to be adopted. For example, although many pollution mitigation measures involve simple best practice which can be implemented without significant cost, many other measures could require substantial investment in pollution prevention technology and/or costly change in land use. Without legislation and/or some form of financial incentive, the likelihood may be low for expensive or complex pollution mitigation measures to be widely taken up on a voluntary basis.

Irrespective of the size of the catchment and local institutional arrangements, water suppliers can often demonstrate a pivotal leadership role in promoting effective catchment protection as shown in Figure 4.4.

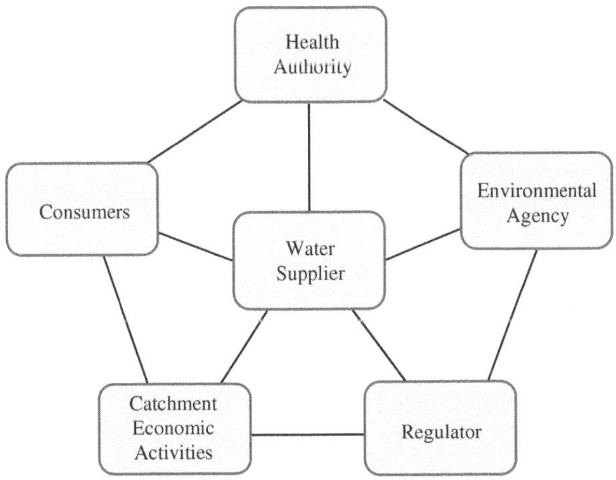

Figure 4.4 Pivotal role of the water supplier in developing catchment partnerships

In working with local catchment partners, a number of different approaches will need to be considered. A key feature of any such partnership will be education and awareness campaigns to alert catchment users to the risks that they can pose to water abstractions and what measures can be taken to reduce those risks. Whatever partnership is initiated, it is important that support and advice are seen as an ongoing process so that land users are actively involved and have a sense of ownership. Approaches which are likely to be most effective will vary considerably, but typically could involve:

- *Obtaining contact details*
  It is important to be able to contact all relevant land users within the catchment who might impact on the water supply abstractions. Depending on the sector involved, this can be time consuming and difficult, particularly in larger or more complex catchments and for diffuse pollution risks.
- *Communication mechanisms*
  This depends on the sector and number of people involved. For small catchments or major land users/premises within larger catchments, personal visits are most likely to be effective. Where larger numbers of people need to be contacted, then personal visits may be impractical and other means such as mail, newsletters and invitations to meetings will need to be considered. It is has been found that holding meetings at the waterworks can help to explain to land users the impact that pollution in the catchment can have on water treatment.
- *Data and information*
  Key to effective communication with land users is provision of relevant information and data including:
  - Local maps showing the water supply catchment in relation to the potential sources of pollution.
  - Data showing the level and variability of pollutants impacting the supply intake.

  - Evidence to show the origin and cause of the pollution.
  - Practical measures that land users can take to reduce pollution or the risk of pollution.
- *Local champions*
  It is often found that farmers and other land users are much more responsive to catchment management messages if they come from people within their own business sector. Thus, seeking to actively engage the support of local business sector "*champions*" who can promote catchment protection messages on behalf of the water supplier can be a very effective approach.
- *Other "influencers"*
  In some cases, land users such as farmers closely follow professional advice from specialist agronomists, or manufacturers/suppliers of agricultural chemicals and machinery. Thus, it is also important to ensure that such "*influencers*" of behaviour are also fully and actively involved in the catchment partnership. Indeed, in some cases their active support may be an important pre-requisite of best practice take-up.

4.2.4 Mapping and characterising the catchment

A crucial early step in catchment management, whether ground or surface, is to ensure that the catchment is properly mapped hydrogeologically. This is usually straightforward for smaller surface water catchments, but for large complex surface water and most groundwater catchments this may require additional specialist expertise. Catchment maps may be in paper format but modern GIS (Geographical Information System) technology now provides a much improved and usually cost-effective option. However, this does depend on the availability of reliable land use, soil type and other spatial datasets.

GIS can be used for electronically interpreting and manipulating the large amount of data necessary for catchment modelling and risk analysis. This can be done by creating maps and displaying a variety of spatial information (e.g. Figure 4.5), allowing visualisation of main catchment characteristics for qualitative and quantitative risk analysis, including physical data (e.g. water bodies, elevation, land use, soil types), the location of monitoring sites and gauging stations, and the location and nature of potential pollutant sources (point and non-point). Spatial variations and temporal trends in water quality conditions can also be very effectively presented and evaluated using a GIS.

Other factors that need to be taken into account in characterising the catchment include the soil type, topography and the nature of surface-groundwater interactions. This allows assessment of the hydrological response to pollution events, natural attenuation mechanisms, travel time from pollution source to abstraction point and the potential impact of extreme weather events such as storms or droughts.

It is particularly important to understand the potential influence that soil type plays in determining the pathways that pollutants take from land to surface and ground waters, and thus the risk of abstraction point contamination. Some soils are prone to rapid transfer of pollutants down preferential flow pathways such as cracks, pipes or animal burrows. For example, some soils with high clay content are prone to cracking in summer and these cracks can persist well into the winter period, acting as conduits for pollutant transport which "*by-pass*" the bulk of the soil matrix. This phenomenon is often made worse by the presence of artificial drainage systems which are designed to remove excess water from the soil, but which also promote rapid land to water transfer of pollutants such as pesticides (Haria et al. 1994). Lighter soils (sandy or loamy textures) tend to be less prone to by-pass flow and pollutants are more likely to travel through the soil matrix.

The nature of unsaturated zone flow (i.e. flow in the soil and rock above the water table) will also influence the risks of groundwater contamination. Where the bulk of water and solute travel via the porous matrix, rather than through large macropores, transport is generally slow and opportunities for contact between water and the solid phase of the medium are many. This encourages sorption of hydrophobic contaminants to any organic matter present (where they can be degraded) and reaction (precipitation, sorption or cation exchange) between inorganic contaminants (such as heavy metals) and the mineral matrix. Aerobic alkaline chemical conditions are normally favourable to significant interception, sorption and elimination of viruses and bacteria.

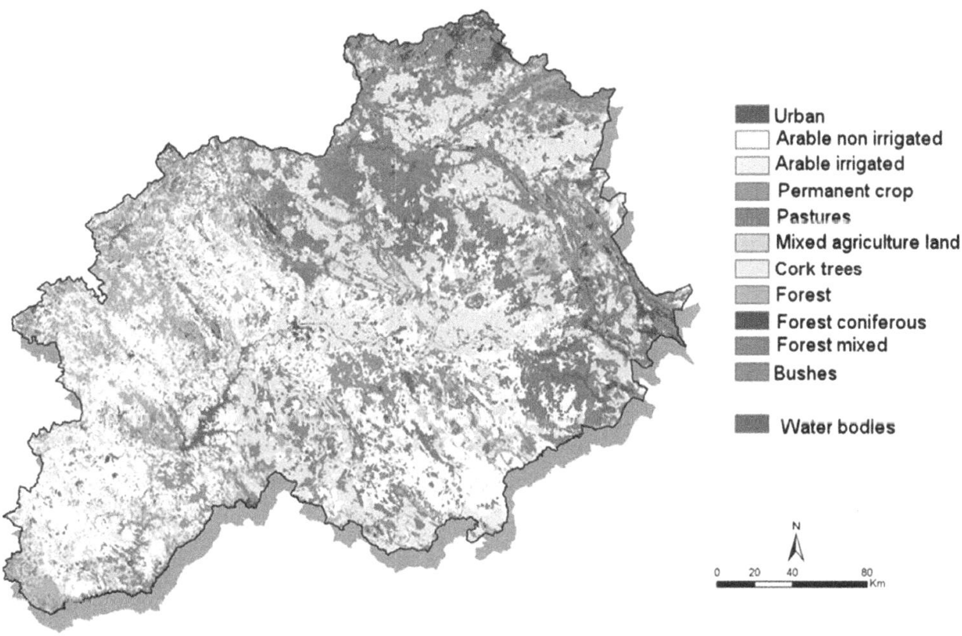

Figure 4.5 Example of a GIS dataset for land use

Soil can, therefore, be the first and most important natural defence against groundwater pollution. Flow velocity plays an important role in these reactions, which is roughly proportional to unsaturated hydraulic conductivity under unit hydraulic gradient and which, for a given soil type, is proportional to moisture content. Below a certain moisture content (known as the field capacity) the rate of flow is very low. All these soil and unsaturated zone properties control the vulnerability of an aquifer to pollution and need to be incorporated in any risk assessment of groundwater.

4.2.5 Identifying hazards and hazardous activities

In conjunction with relevant catchment stakeholders, potential sources of pollution impacting the abstraction point should be identified and marked on the catchment map, using standard symbols. These may be both point source (e.g. sewage works, factory, other commercial premises) or diffuse source (e.g. intensive farming, roads and railways, urban areas, oil pipelines). More information on detailed methodologies for assessing and managing surface water is given in Section 4.3 and groundwater in Section 4.4.

It is important to recognise that extreme weather events can influence both the quantity and quality of surface and groundwater. The hydrological regime in a catchment (i.e. the timing and typical rates of infiltration, overland flow, drainflow, groundwater recharge and stream flow) depends on the seasonality and rate of precipitation and evapotranspiration. Extreme rainfall and water shortage during drought periods have significant impacts on river and spring flows, lake and reservoir levels, and the depth of the groundwater water table. These in turn can impact raw water quality.

For example, surface water quality visibly deteriorates after extreme rainfall, when increased erosion and in-stream turbulence result in elevated suspended solids concentrations (turbidity). During drought, higher pollutant concentrations may occur due to a reduction in the dilution capacity of receiving waters. Eutrophication in source waters can be greatly stimulated by high temperatures.

Although groundwater is normally less susceptible to rapid change due to climatic effects, extreme weather events may nevertheless cause impact on aquifers. Long drought periods will reduce groundwater recharge, thus decreasing contaminant dilution and water availability. Intense rainfall events are likely to result in a greater fraction of precipitation being transferred to surface waters via rapid pathways, rather than recharging groundwater, and may therefore exacerbate the effects of extreme droughts on water table levels. Flooding caused by such events may also drive contaminated water into the aquifer through wells.

4.2.6 Assessing risks, promoting mitigation measures and verifying their effectiveness

The hazards and hazardous activities identified in the catchment should be assessed and prioritised in terms of likely risk at the point of abstraction. In so doing it is particularly important to take account, not just of those potential pollutants which could have a rapid impact on the waterworks, for example chemical spillage, but those which could deteriorate raw water quality over a long time period if mitigation measures are not implemented, such as groundwater pollution from nitrate (Howden et al. 2011). To support such risk assessment and identify effective mitigation measures, it is best to develop a source/pathway/receptor model. This identifies the source of any potential pollution, the pathways through which it travels and the receptors (e.g. water abstractions) which might be impacted.

Having identified the priority risks to water quality, catchment risk mitigation measures should be promoted in conjunction with the relevant catchment management authorities and potential polluters. More information on common catchment risks, their mitigation and potential control points to assess their effectiveness, are given in Section 4.5 later in this Chapter.

4.2.7 Implementing risk-based raw water monitoring

4.2.7.1 Introduction

An important part of any catchment management strategy includes implementation of raw water monitoring programmes, targeted at the most likely parameters of concern and based on catchment characterisation and associated risk assessment. Monitoring programmes can, however, be very expensive to set up and maintain. Thus, if the sampling design can be targeted to key determinands, with frequency adjusted at specific times of the year, greater efficiency can be obtained. For example, knowledge of the scale and condition of sewerage and sewage treatment facilities, as well as the location and intensity of livestock farming in a catchment, can help to identify the extent of likely microbiological challenges to water treatment. This particularly needs to take into account the fact that the microbial load can be considerably higher during storm periods or snow melt.

Similarly, knowledge of the scale and type of farming activity in the catchment, particularly when linked to catchment fate and behaviour models, can help to predict the occurrence of agricultural chemical contaminants such as nitrate or pesticides. This includes the prediction of high episodic peaks which might be missed by infrequent spot sampling. This can be done by employing a model of diffuse source pollutant exposure for the upstream catchment. If the model can be validated to demonstrate that it is a reliable predictor of exposure, then monitoring can be effectively targeted only at those parameters likely to be of concern and at a time and frequency when they are most likely to occur. Thus, the monitoring frequency could be increased during periods of high risk in order to help to identify the worst case challenges to water treatment. Conversely, at other times of the year when such chemicals are not used, sampling frequency can be scaled back.

Groundwater sampling programmes present additional challenges. The conceptual flow model of the aquifer has to be understood prior to developing the sampling strategies. Generally slower travel times and a greater degree of mixing compared to surface water, along with natural attenuation characteristics, make both the sampling frequency and the parameters of interest different to those used for surface water resources. More information is given later in Section 4.4.7.

4.2.7.2 Objectives of monitoring

Monitoring can have a number of objectives:

- *Sampling for design and operation of treatment plant*
  Routine risk-based monitoring programmes can help to build up a picture of expected raw water quality and the extent of any quality variability. This allows the treatment plant to be designed and operated with sufficient spare capacity to deal adequately with worst predicted raw water quality (See Chapter 5).
- *Sampling to detect raw water quality deterioration*
  Even appropriately designed and operated treatment plants can be subject to risks caused by unexpected deterioration in raw water quality. Such deterioration can occur due to accidental pollution, introduction of new industrial activities, changed agricultural practices or extreme climatic events. Knowledge of land use or activities in the catchment which might pose a risk to water quality facilitates the design of monitoring systems that can provide early warning of raw water quality deterioration. However, although routine spot sampling and analysis can be very useful for detecting quality deterioration which occurs over a matter of days or longer, it cannot usually detect very rapid deterioration of raw water quality or episodic peaks. This is particularly the case where samples are sent to an external laboratory for analysis, because there will be delay in reporting which can impede rapid response to pollution incidents (Whelan et al. 2011). However, such quality alteration might be picked up by automatic intake monitoring systems. For groundwater, except in aquifers of extreme or high pollution vulnerability, it will not normally be necessary to monitor aquifer groundwater quality more frequently than, for example, three month intervals.
- *Sampling as part of catchment management*
  Raw water sampling is also important to supporting catchment management plans as part of the risk assessment process. It allows a better understanding of the background (baseline) natural water condition, thus allowing long term trend assessment for parameters of concern. It can also be used as a way of monitoring the effectiveness of catchment risk mitigation measures.

4.2.7.3 Sampling programme design

The cost effective design of raw water sampling programmes can only be determined by the water supplier based on a good knowledge of the raw water catchment, and taking into account available resources, skills and laboratory capability. The programme should be formally reviewed at least once a year, or following an event or incident in the catchment which might impact on raw water quality. Approaches might include:

- *Initial sampling programme*
  Before bringing a new source into supply, sampling will need to take place for a period and at a frequency suitable for the type of source. The purpose of this sampling is to establish any quality risks and the necessary level of treatment. For groundwater it is also important to evaluate other existing wells/springs in the area to detect water quality problems before the construction of a new well. In the case of natural contamination risks, it is necessary to also evaluate the presence of substances which could leach into the saturated zone as a result of changing redox or pH conditions which are common as a result of groundwater development.
- *Core routine programme*
  A basic suite of physicochemical and microbiological analysis will need to be put in place, geared to the type of source, its variability and the level of treatment in place. In some cases more frequent monitoring for simple indicator parameters may be used as a cost effective complement to other determinands which are more difficult or expensive to measure. For example, ammonia might be used as an early warning of increased levels of sewage or animal wastes. In the case of groundwater, both nitrate and chloride are very common indicators of aquifer quality, due to their chemical stability and mobility in subsurface environment. If the catchment characterisation suggests that the raw water quality will be stable (e.g. deep groundwater) then the monitoring frequency required may be quite low. Conversely, if the groundwater source is shallow or if the rock is karstic, then a higher frequency of sampling will be required. A surface water sampling programme will, of necessity, involve a wider range of parameters and at a higher frequency to reflect the greater range of risks and expected increase in spatial and temporal variability.
- *Event related sampling*
  The core routine sampling programme should be supplemented by additional sampling at certain times of the year. This could be when quality would be expected to deteriorate due, for example, to storm events, drought or when additional catchment risks occur (e.g. use of pesticides for crop protection, additional strain on sewerage infrastructure due to tourist influx).
- *Long time series sampling for trend analysis*
  Some catchments might be subject to long term deterioration over a period of years due, for example, to changes in land use or agricultural practices. In such cases it is important that monitoring maintains adequate long time series data for key parameters such as nitrate or salinity (Burt et al. 2008 and 2010). Short term cost savings should not impact the collection of such data.

4.2.7.4 Automatic intake monitoring systems

Reliable and continuous monitoring systems have been increasingly used in routine long-term monitoring, as well as in early warning of accidental or malicious source water contamination. Initially, water quality monitoring systems were limited to easily measurable physicochemical parameters such as temperature, pH, conductivity and dissolved oxygen. Nowadays, it has evolved to include sophisticated equipment that allows monitoring for a range of parameters including organic pollutants. For example, equipment used in the UK on some surface water intakes employs on-line solid-phase extraction followed by liquid chromatography and can generate frequent analyses which can be made available almost immediately (Whelan et al. 2011). Other alternatives to conventional analysis include ELISA (Enzyme-linked Immunosorbent Assay) kits, which rely on substance specific binding of antibodies to the target compound which are then detected spectro-photometrically. Although these kits can be relatively expensive, they do not require investment in expensive instrumentation or dispatch of samples to commercial laboratories, so allowing rapid determination of concentrations and subsequent communication.

Increasingly, novel continuous bio-monitoring techniques can also be used to detect potentially toxic contaminants in source waters. These rely on living organisms such as daphnids, algae, and fish as indicators of a range of potentially toxic substances, although these need to be calibrated to ensure they are reacting to those substances which are of concern to drinking water quality rather than ecological impact.

4.2.8 Implementing catchment warning and response procedures

Pollution incidents potentially impacting water supply abstractions will occur from time to time, even where effective arrangements for catchment management and control have been implemented. These may not be detected by intake monitoring systems and, even if they do, it may not allow sufficient time for appropriate response. Thus, effective communication systems between water suppliers and catchment managers/users are important to give early warning of pollution events. Whilst pollution incidents are most likely to impact surface water catchments, particularly those that are large and/or complex, pollution incidents can also impact on groundwater abstractions. Both should therefore be considered for implementation of pollution warning systems.

4.2.8.1 Organisations typically involved

As well as the water supplier, the organisations that need to be involved in a pollution warning system will depend on local catchment characteristics, legal and institutional arrangements and the catchment risk assessment. Particular complications can arise where catchments cross state or national boundaries. Organisations could therefore include:

- Water utility staff.
- Catchment management agencies.
- Municipal authorities including emergency services (fire, police).
- Other upstream abstractors.
- Catchment users such as major industrial sites, farmers, fishing organisations.

In addition it should be recognised that members of the general public might also report possible pollution incidents or unusual catchment events which need to be rapidly and effectively captured, assessed and communicated.

4.2.8.2 Early warning procedures

The early warning procedures that need to be put in place will vary, depending on the local situation and available communication systems. Whatever procedure is developed, it will need to be simple, agreed by all parties and clearly documented. Once implemented the procedures should be regularly reviewed to identify areas for improvement. Consideration should also be given to include pollution warning systems in any exercises held to test the effectiveness of incident management procedures (see Chapter 3.6).

4.2.8.3 Pollution awareness training

Catchment managers and users often do not appreciate the range of pollutants that might impact water supply intakes. As well as toxic material this could include:

- Substances which might impart a taste to water supplies, even at low levels, such as oils, solvents or phenolic compounds.
- Substances which might impact on disinfection effectiveness by exerting increased oxidation demand.
- Events such as fires at industrial premises, major transport accidents, fuel or chemical spills, natural events such as floods or landslip.

For this reason, training and awareness programmes should be considered for key external personnel. This should include:

- The types of pollutant or pollution event that should be notified.
- Why these might impact on water supply intakes and treatment.
- The importance of speed of notification.
- Information which should be captured at time of notification.

4.2.8.4 Effective, timely and accurate communication

Key to effective warning systems is a rapid, foolproof and clearly defined communication mechanism between those identifying the event and the relevant water supplier staff. Such a system could have different levels of warning depending on the severity of pollution risk (e.g. Level 1 - early warning; Level 2 - confirmed risk; Level 3 - serious and imminent risk). Once a pollution incident has been resolved, an "*all clear*" message should be sent.

The communication method used will depend on local circumstances, but to avoid mistakes or misunderstanding the method should preferably be captured on a standard form which is transmitted electronically e.g. via fax or e-mail. The recipient of the warning message within the water supplier should be identified and procedures should ensure that it is picked up and responded to promptly 24/7. Normally, this might be at a continuously manned control room or duty officer, but other arrangements may also be appropriate.

For groundwater, the rate at which contaminant plumes travel is usually orders of magnitude lower than in surface waters. Therefore, it is not normally necessary to have 24/7 response teams, except for karstic and high transmissivity fractured rock aquifers.

4.2.8.5 Pollution travel time estimation

To improve effective response to pollution warning it is important to be able to assess the potential travel time from the site of the pollution incident to the water abstraction point. These are defined by the travel distances and intrinsic characteristics of each environment:

- Rivers: the distance involved, the discharge and velocity and the behaviour of the pollutant concerned.
- Lakes and reservoirs: the residence time of the system including wind effects and short circuiting.
- Groundwater: the distance is defined by the thickness of the unsaturated zone, and the location of source and abstraction points, which will roughly determine the travel distance in the saturated

zone. Pollutant characteristics must also be taken into account. Fissured aquifers, or where natural protective zones are bypassed, are particularly vulnerable.

In larger, complex or more vulnerable catchments it may be appropriate to develop practical computer models to forecast time of travel based on the hydrological properties of the water resource system and pollutant behaviour.

4.3 METHODOLOGIES FOR ASSESSING RISKS AND MANAGING SURFACE WATER CATCHMENTS

4.3.1 Introduction

Integrated management of surface catchments must identify all the existing and potential hazards that might pose a risk to surface water abstractions and the pathways through which the hazard might travel to the water abstraction point. This, in turn, can support the implementation of effective source water protection programmes.

A full risk assessment must include the study of ecological and hydrological processes governing catchment systems, as well as anthropogenic polluting activities. The identification and characterisation of pollutant sources are critical to the successful mapping of risks in surface catchments (CCME 2004). Table 4.1 shows some major factors affecting source water quality.

Table 4.1 Factors affecting source water quality

Natural Factors	Human Factors	
	Pollutant non-point sources	**Pollutant point sources**
Climate	Agricultural cropland run-off	Industrial discharges
Topography	Livestock/grazing	Wastewater discharges
Geology	Dairies and feedlots	Hazardous waste facilities
Soil cover	Urban development runoff	Mine drainage
Vegetation	Septic tanks	Spills and releases
Fire	Erosion	Urban runoff
Wildlife	Forest management	Combined sewer overflows
Saltwater Intrusion	Mining	Aquaculture
Thermal stratification	Recreational activities	
Erosion	Atmospheric deposition	

4.3.2 Use of predictive models

Predictive surface water models provide a very good approach for evaluating alternative catchment management scenarios. When combined with good monitoring datasets, properly calibrated, tested and verified models can be used to forecast or estimate risks under various scenarios. This allows a good insight into impacts associated with known and anticipated land use activities within the catchment. Risk-based decision-making should be based on a consistent interpretation of both current understanding and the available data. A tiered approach to modelling ensures that the optimum effort is expended for each environmental decision.

Different type of models can be applied to simulate and predict surface water quality (USEPA 2008). The decision-making process in catchment planning and management is multidisciplinary in nature and integrates different sources of knowledge (scientific, socioeconomic and political). Decision support systems attempt to combine complex, dynamic knowledge from a number of disciplines in a user-friendly graphical user interface. They can be used to organise information and to design and assess the impact of alternative long-term catchment management plans. Figure 4.6 depicts the structure of a web-based decision support system, developed for supporting integrated water management of the Portuguese River Guadiana basin (Vieira et al. 2008).

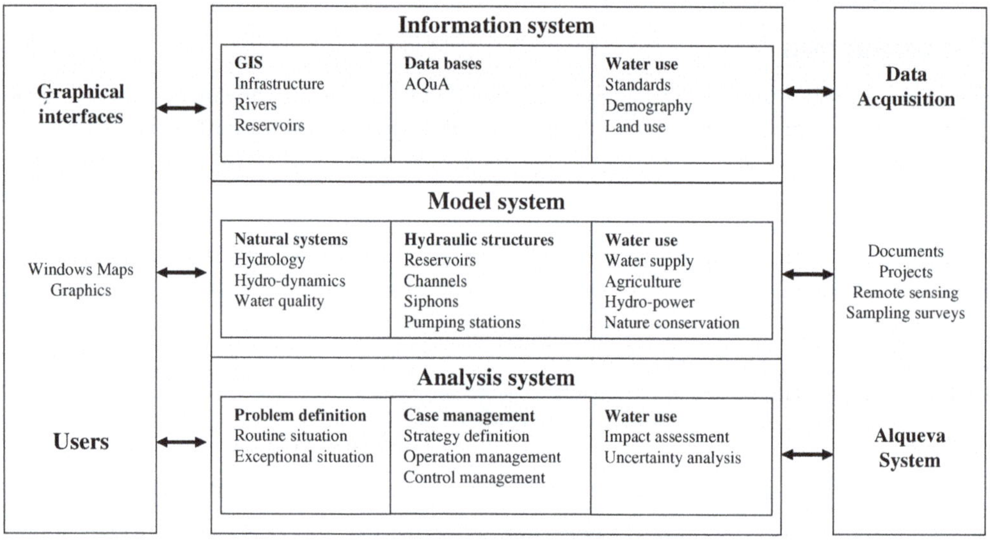

Figure 4.6 Decision support system structure in River Guadiana basin

4.4 METHODOLOGIES FOR ASSESSING AND MANAGING GROUNDWATER CATCHMENTS

4.4.1 Strategies for control of groundwater pollution

Implementing groundwater pollution prevention measures is a complex task, involving two interrelated but independent strategies which focus on the protection of:

- Groundwater resources (aquifers) as a whole and/or
- Groundwater sources - those parts of the aquifers where the resource is exploited for water supply or other purposes.

While both approaches are complementary, the emphasis placed on one or other will depend on local hydrogeological conditions, the extent to which groundwater resources have been developed and exploited, and a range of other broader socioeconomic factors.

Once sub-surface or groundwater pollution has occurred it is usually very expensive and sometimes impractical to implement remedial action at a reasonable economic cost. The exception may be those fast response aquifers where removal of the hazard or hazardous activity may result in reduction of groundwater contamination in a relatively short timescale. The primary aim of groundwater protection should therefore be focussed on prevention of those activities which might result in unacceptable sub-surface contamination either across the whole aquifer or within the capture zone for public supply sources.

There are four main catchment protection tools (see later sections for more detail) which can be used to control groundwater pollution risks:

- Mapping overall aquifer pollution vulnerability.
- Quantifying the potential sub-surface contaminant load.
- Implementing Source Protection Areas (SPAs) which define the zone of capture of boreholes, wells and springs.
- Placing restrictions on hazards or hazardous activities within defined areas of the catchment.

In developing a WSP, ideally the whole aquifer should be considered for protection. However, depending on regulatory policies and other local factors, in some situations it may not be economic or practicable to do so. For example, it may not be cost effective to protect the whole aquifer if only a small amount of it is used as water source or for other purposes. In such situations, it would suffice to define the groundwater capture zone of specific water sources, and assess their pollution vulnerability and sub-surface contaminant load, usually on a scale of 1:10,000 to 1:50,000. Conversely, in those areas with a large number of significant groundwater abstractions, whole aquifer strategies should be adopted.

The responsibility for implementing each strategy may vary. Aquifer pollution vulnerability assessment is a management tool used as part of land use planning to protect strategic aquifers. Depending on the size of the aquifer, these are, therefore, usually the responsibility of regional or national authorities. Source pollution hazard assessment, on the other hand, is more useful for local catchment managers and water suppliers.

Whole aquifer strategies are more universally applicable, since they endeavour to achieve a degree of protection for the entire groundwater resource and for all groundwater users. They would commence with aquifer pollution vulnerability mapping of more extensive areas (including one or more important aquifers) working at a scale of 1:100,000 or more, if the interest was limited to general information and planning purposes. This would be followed by the identification, localisation and classification of potential contaminant loads. The interaction between these two elements permits the aquifer pollution hazard definition (see Figure 4.7 below).

The same approach can be followed for source protection, where the pollution hazard is established through superimposition of the outputs from the sub-surface contaminant load inventory on the different well capture zones, as set out in Figure 4.8 below.

4.4.2 Objectives for groundwater pollution prevention

There are three different scenarios to be considered in establishing a groundwater quality protection strategy.

4.4.2.1 Preventing Future Pollution

Aquifer pollution vulnerability maps are a valuable tool to reduce the risk of creating future groundwater pollution hazards. They identify the areas most vulnerable to groundwater pollution, so that the location of potentially hazardous activities can be avoided or prohibited (see Section 4.4.6 later).

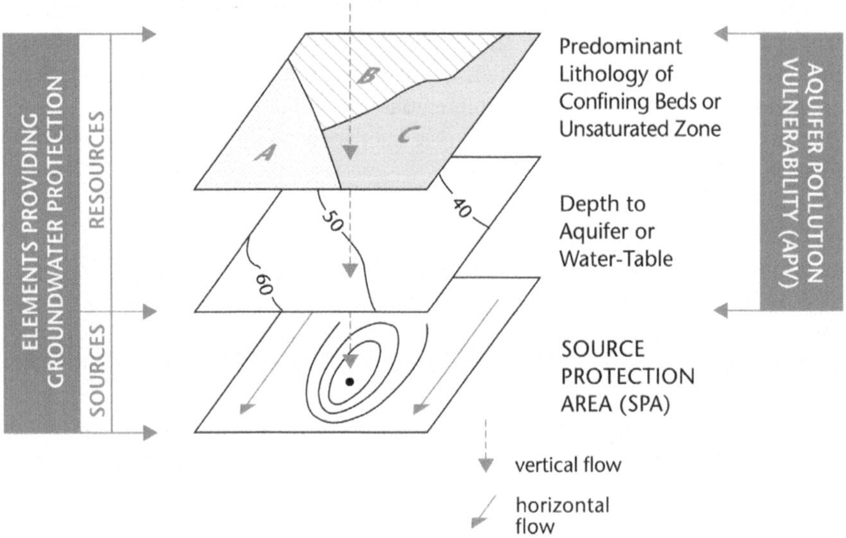

Figure 4.7 Components of groundwater pollution hazard assessment used for land surface zoning (Foster et al. 2002)

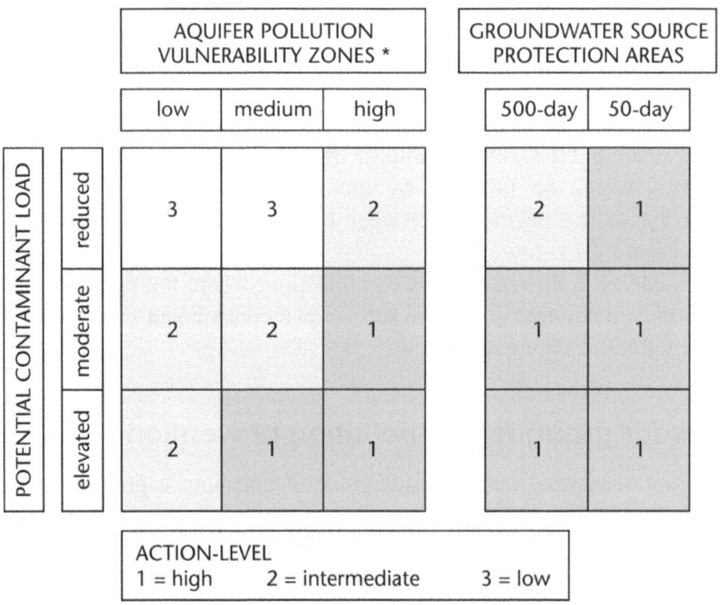

* Numbers of zones/areas reduced to simplify presentation.

Figure 4.8 Priority groundwater pollution control action-levels based on aquifer vulnerability, source protection areas, and potential contaminant load (Foster et al 2002)

4.4.2.2 *Dealing with Existing Pollution Sources*

In this situation, it is necessary to prioritise groundwater pollution control measures in areas where a range of potentially polluting activities already exist. First, it will be necessary to establish which among these activities poses the most serious hazard to groundwater quality, using aquifer vulnerability mapping, delineation of water supply protection areas, and inventories of the sub-surface contaminant load (see below).

4.4.2.3 *Selecting New Groundwater Supply Areas*

In this situation, it is first necessary to define the capture zone of any new public supply well and then to identify all potential contaminant sources existing in the area (see Section 4.4.5 later). Where such an assessment identifies anthropogenic activities capable of generating an elevated sub-surface contaminant load, and/or the aquifer pollution vulnerability is high or extreme over most of the designated groundwater supply capture area, this assessment should be followed by a technical and economic appraisal to establish the possibility of controlling all potential pollution activities. If this is not possible then alternative sites for new groundwater supply sources should be investigated.

4.4.3 Mapping overall groundwater pollution vulnerability

The concept of aquifer vulnerability is related to the intrinsic characteristics of the strata which separate the saturated aquifer from the land surface. These characteristics determine the risk of the aquifer being adversely affected by a surface-applied contaminant load (Foster & Hirata 1988). Vulnerability is determined by:

- A hydraulic assessment of likelihood that the pollutants will reach the saturated aquifer and
- An estimate of the extent to which pollutants will be attenuated by physiochemical retention or reaction during passage through the strata overlying the saturated zone.

Aquifer pollution vulnerability maps (Holman et al. 2005) are designed to provide a general framework within which groundwater protection policy can be developed. They are a simplified, but factual, representation of the best available scientific data on the hydrogeological environment. Generally, the methods provide an overall index for a single integrated vulnerability to all potential pollutants. However, some methods allow aquifer vulnerability to be determined for individual contaminants, groups of contaminants or even for activities or particular pollution scenarios. It should be noted though, that vulnerability maps for specific contaminants and pollution scenarios tend to be much more costly, complex and difficult to use than generic maps.

There are many different aquifer pollution vulnerability methods. One example is known as GOD (Foster and Hirata, 1988) which defines an absolute index, unlike other methods that provide only relative vulnerability indices. It uses factors for three intrinsic geologic and hydrogeological characteristics:

- Groundwater hydraulic confinement in the area under consideration.
- Overlying strata, in terms of character and degree of consolidation that determine their contaminant attenuation capacity.
- Depth to groundwater table.

There may also be locally available groundwater pollution vulnerability methodologies used by regulatory authorities or other catchment management agencies (e.g. UK Environment Agency 1999).

There are a number of hydrogeological conditions that create problems for vulnerability assessment and mapping, and this means that expert support and interpretation may be necessary. Also, the methods used to generate vulnerability maps do not cover two specific situations: contaminants discharged directly in the sub-surface, e.g. from underground storage tanks or septic systems, and spillages of dense non-aqueous phase liquids (DNAPLs) such as chlorinated solvents used for dry cleaning. A high groundwater pollution hazard is present in both situations, regardless of aquifer vulnerability.

4.4.4 Inventory and classification of subsurface potential contaminant load

A key step in implementing a groundwater protection strategy is to undertake a detailed survey of the aquifer catchment or source protection area to identify those hazards or hazardous activities which have the potential to cause groundwater pollution. In doing so, it is important to take into account whether generation of a sub-surface contamination load is a direct result of the hazard, for example from septic tanks, or whether the load might be generated accidentally, for example due to spillage (Foster et al. 1993). It is also important to understand whether any hazardous activities have taken place in the past, since polluting processes or activities which ceased some years ago can still generate a sub-surface contaminant load by leaching from contaminated soil. Similarly, some pollutants derived from past agricultural activity may take many years, or even decades, to reach the aquifer or borehole due to long unsaturated zone travel times (Howden et al. 2011).

There are various published methods of assessing the pollution potential of human activities, although few are directed specifically to their potential to generate a sub-surface contaminant load (Foster and Hirata 1988; Johansson and Hirata 2001). The classification of potentially polluting activities by their spatial distribution provides a direct and visual impression of the type of groundwater contamination threat they pose and the control measures which are likely to be required:

- Single substance point pollution sources normally cause clearly defined and more concentrated plumes, which makes their identification (and in some cases control) easier. However, this does not always occur with complex mixtures of substances lost, for example from industrial sites and/or in complex layered aquifers. Additionally, when multiple point-source pollution activities are small they essentially act in the same way as diffuse sources, with respect to identification and control.
- Diffuse pollution sources do not generate clearly defined groundwater pollution plumes, but they normally impact a much larger area (and thus volume) of aquifer.

Using any of these classifications, it is always useful to rank the potential contaminant load in terms of the probability to generate a load that can reach and impact the saturated aquifer. There are some methods that classify the potential load in three levels: elevated, moderate and reduced, for example the POSH method (Foster et al. 2002). This is based on two intrinsic characteristics:

- The likelihood of the presence of contaminants, which are known or expected to be persistent and mobile in the sub-surface; and
- The existence of an associated hydraulic load (surcharge) capable of transporting contaminants into and through an aquifer.

4.4.5 Groundwater source protection areas (SPAs)

The delineation of groundwater source protection areas (SPAs) or zones (SPZs) is an essential element in the protection of drinking water sources from contamination. SPAs have to take into account two different types of contaminant:

- Those that decay with time, where sub-surface residence time is the best measure of protection.
- Non-degradable contaminants, where attenuation of concentration depends on dilution and hydrodynamic dispersion.

The proximity of any potentially polluting activity to a groundwater source is a key factor influencing the contamination risk it poses. The pollution threat will depend on whether the activity is located within the capture zone of the water supply source, and the horizontal flow time in the saturated aquifer from the location of the activity to the abstraction point (Figure 4.9). Source protection areas are, therefore, normally divided into four zones:

- Zone A - The operational area immediately adjacent to the wellhead or borehole.
- Zone B - The inner catchment area representing approximately 50 day travel time (in the saturated zone) to the abstraction point which is normally considered the minimum necessary for degradation of microbiological pollutants.
- Zone C - the intermediate area which, depending on the aquifer, represents approximately 1–10 years travel time to the abstraction point, and which allows time for remedial action to take place if problems occur.
- Zone D - the whole source capture zone.

There are five main approaches used for delineating groundwater SPAs (USEPA 1994), in increasing order of complexity and cost:

(1) Arbitrary fixed/calculated radius around borehole.
(2) Simplified variable shapes.
(3) Analytical hydrogeological models.
(4) Hydrogeological mapping.
(5) Numerical groundwater flow modelling with particle tracking.

Arbitrary fixed-radius circular zones (1) or simple elliptical shapes (2) have commonly been used. However, because of their questionable reliability they would normally be chosen only in the very first stage of groundwater source protection programme. The choice of other methods to employ will depend on availability of hydrogeological data and available resources. However, the geometry of the protection zone defined will also be influenced by the method used for its delineation. It must also be remembered that the delineation of protection zones is a dynamic process. Groundwater conditions may physically change or new hydrogeological data may come to light that enable the aquifer to be more accurately represented. Given the numerous uncertainties and continuously evolving groundwater conditions, it is wise for an *"adaptive approach"* to SPA implementation strategy. This should be based on the application of analytical method in the first instance, followed by a numerical method when more data are available.

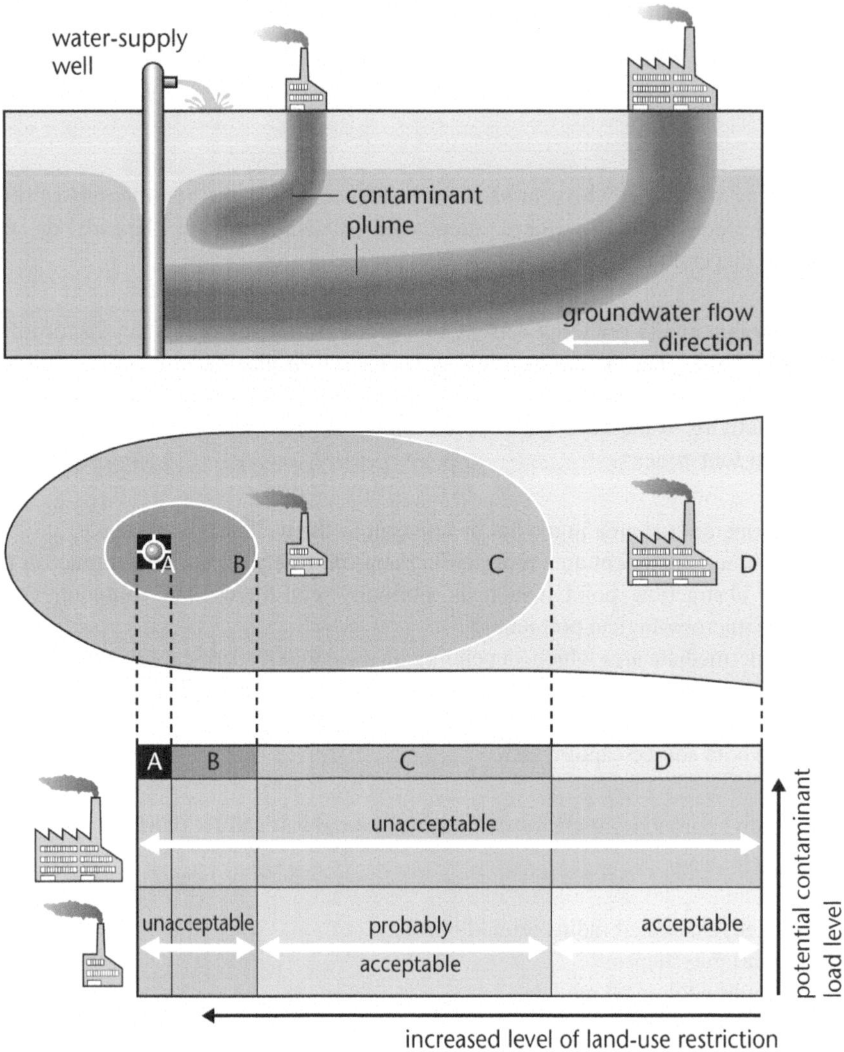

Figure 4.9 Concept of different levels of source protection areas, with land use restrictions (Foster et al. 2002)

4.4.6 Placing restrictions on hazards or hazardous activities within defined catchment areas

As previously described, the key to effective groundwater protection is to identify the vulnerability of either the whole aquifer or the catchment of a water supply source to sub-surface contamination and then control the hazard or hazardous activity which might give rise to such a polluting load. Table 4.2 below, sets out a typical acceptability matrix for various hazards according to aquifer vulnerability. The terminology used is:

- N = unacceptable in virtually all cases.
- PN = probably unacceptable, except in some cases subject to detailed investigation and special design.

- PA = probably acceptable subject to specific investigation and design.
- A = acceptable subject to standard design.

Table 4.2 Acceptability matrix of common potentially polluting activities and installations according to aquifer vulnerability

Potentially polluting activity requiring control	Aquifer vulnerability		
	High	**Medium**	**Low**
Septic tank, cesspits and latrines			
Individual properties, communal properties, public facilities	A	A	A
Gasoline station	PA	A	A
Solid waste disposal facilities			
Construction/inert	A	A	A
Municipal domestic and industrial (class I)	PN	PA	A
Industrial (class II and III) and hazardous	N	N	PA
Cemetery	PA	A	A
Incinerator	N	PN	PA
Mineral and oil extraction			
Construction material (inert)	PA	PA	A
Others, including petroleum and gas	N	PA	A
Fuel lines	N	PA	A
Industrial premises			
Type I	PA	PA	A
Type II and III	PN/N	PA/N	PA/PN
Infiltration lagoons			
Municipal/cooling water	A	A	A
Industrial effluent	PN	PA	PA
Soakaway drainage			
Building roof	A	A	A
Major road, industrial sites, airport/railway station	PN	PA	A
Minor road, parking lots	PA	A	A
Effluent land application			
Food industry, sewage effluent and sewage sludge	PA	A	A
All other industries	PN	PA	A
Agricultural areas			
Land application of farmyard slurry	A	A	A
Intensive livestock effluent lagoon, farmyard and feedlot drainage	PA	A	A
Arable use of pesticides	PN	A	A
Pesticide storage	PN	PA	A
Arable uncontrolled use of fertilizers	PN	A	A

Table 4.3 below, sets out a typical acceptability matrix for various hazards, using the same criteria, within different source protection zones as described in Section 4.4.5. In summary these are:

- A: Operational area.
- B: Microbiological protection area.
- C: Intermediate protection area.
- D: Entire capture area.

Table 4.3 Acceptability matrix of common potentially polluting activities and installations according to source protection areas A, B, C, D

Potentially polluting activity requiring control measures	Source protection area			
	A	B	C	D
Septic tanks, cesspits and latrines				
Individual properties	N	N	A	A
Communal properties, public facilities	N	N	PA	A
Gasoline station	N	N	PN	PA
Solid waste disposal facilities				
Municipal domestic	N	N	N	PN
Construction/inert	N	N	PA	PA
Industrial hazardous	N	N	N	N
Industrial (class I)	N	N	N	PN
Industrial (class II and III)	N	N	N	N
Cemetery	N	N	PN	A
Incinerator	N	N	N	PN
Mineral extraction				
Construction material (inert)	N	N	PN	PA
Others, including petroleum and gas	N	N	N	N
Fuel lines	N	N	N	PN
Industrial premises				
Type I	N	N	PN	PA
Type II and III	N	N	N	N
Military Facilities	N	N	N	N
Infiltration lagoons				
Municipal/cooling water	N	N	PA	A
Industrial effluent	N	N	N	N
Soakaway drainage				
Building roof	PA	A	A	A
Major road	N	N	N	PN
Minor road	N	PN	PA	PA
Amenity areas	N	PA	PA	A
Parking lots	N	N	PN	PA
Industrial sites	N	N	N	PN
Airport/railway station	N	N	N	PN

4.4.7 Groundwater monitoring strategies

Groundwater monitoring strategies include the choice of using existing production wells, or drilling dedicated monitoring wells (Figure 4.10). Using existing production wells might initially be appealing due to the lower cost and implementation time, but poses serious drawbacks. Production wells usually have large intake screen intervals, and mix waters from widely different depths, with different residence times and hydrochemical properties. Sampling from a wellhead tap, usually close to high capacity pumping plant, may also impact sample quality by allowing contact with air and causing oxidation and precipitation of Eh-sensitive constituents, volatilisation and modification of pH. Purposed-drilled monitoring wells need to be carefully placed, taking account of the conceptual flow model of the aquifer and the objective of the monitoring programme. In a more complex aquifer system where vertical water flow is expected, multi-level monitoring wells have to be considered. There are three strategies that can be adopted for groundwater monitoring:

- Offensive (proactive) monitoring of potential pollution sources, where the objective is to provide early detection of aquifer contamination from known sources of potential pollution. Monitoring is undertaken immediately down the hydraulic gradient, and analytical parameters are chosen specifically in relation to the pollution source. This approach is expensive and, thus, has to be highly selective, primarily targeting the more hazardous pollution sources located within groundwater supply capture zones in aquifers of high-moderate pollution vulnerability.
- Defensive monitoring for groundwater supply sources, when the objective is to provide warning of pollution plumes threatening potable well fields or individual water wells and springs. This is achieved through the installation of a monitoring network up the hydraulic gradient, which is capable of detecting approaching polluted groundwater in time for further investigation and remedial action to be taken. A thorough understanding of the local groundwater flow system and contaminant transport pathways is required, especially in relation to selection of the depths of monitoring borehole intakes, to avoid the possibility of by-pass of the defensive monitoring network.
- Evaluation monitoring for sites of known aquifer contamination. A similar approach to that described under proactive monitoring should be adopted. This serves two purposes: (1) most importantly to confirm the effectiveness of natural contaminant attenuation processes, where these are considered to be the most economic or only feasible way to manage aquifer pollution, and (2) to confirm the effectiveness of remedial engineering measures taken to clean up or contain aquifer contamination, where these have been judged technically and economically feasible.

4.5 COMMON CATCHMENT RISKS AND THEIR MITIGATION

4.5.1 Introduction

Previous sections of this chapter have described factors underpinning catchment WSPs, approaches to developing a catchment WSP, and methodologies for managing both surface and groundwater catchments. This section briefly summarises a number of common catchment hazards which might pose a risk to water treatment works and typical approaches to their mitigation. More information is given in references on microbial hazards (WHO 2011) and potential chemical hazards (WHO 2011, Thompson et al. 2007).

Effective pollution control in catchments can be secured through a wide range of policy approaches. These can include legislation, financial or commercial instruments, technical guidance and voluntary codes of best practice. Many countries use a combination of all these measures, with pollution control legislation complemented by other fiscal and voluntary approaches. However, it has to be remembered that legislation and other measures are only effective if coupled with appropriate levels of enforcement and inspection at a local level.

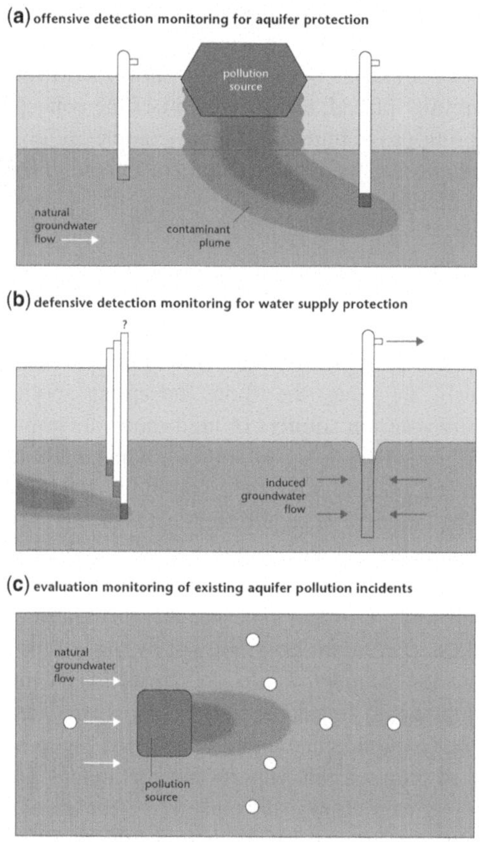

Figure 4.10 Schematic summary of groundwater quality monitoring strategies (Foster et al 2002)

Once mitigation measures have been implemented it is important to verify the extent to which these are effective. This is normally the responsibility of the catchment control authorities, but water suppliers will need to work closely with such organisations to ensure that the pollution control regime adopted is effective in protecting their abstractions.

The control points for different catchment risks will vary considerably but could include:

- Raw water monitoring within the catchment or at the point of abstraction.
- Routine monitoring and inspection of point source discharges as well as industrial, commercial or other high risk sites.
- Routine inspection of land use to confirm whether or not any required changes in practice have occurred (e.g. livestock density, changed land use or cropping).

- Maintenance of records by land users of such things as pesticide and fertiliser usage.
- Training and certification requirements for land users.

The particular pollutants that need to be considered will vary widely from catchment to catchment, but could include one or more of the following:

4.5.2 Naturally occurring surface water hazards

There are a number of pollution hazards which can arise naturally in surface waters. However, in many cases these can be made worse by human activities. They include:

- Metals such as iron, manganese and aluminium which, if not removed by treatment, can give rise to water discolouration.
- Increased turbidity, particularly after rainfall which can impact the efficiency of disinfection and other treatment processes.
- Natural organic matter (NOM), which can result in discolouration, adversely impact disinfection effectiveness (for example, through increased oxidation demand), and also lead to increased formation of disinfection byproducts such as THMs (Lewin et al. 2004). Depending on the composition it can also provide a nutrient source for microbial regrowth within the distribution network.
- Nutrients such as nitrogen and phosphorus which are naturally present but which can be augmented by agricultural activity and waste water discharge. This can lead to the proliferation of algae in the phenomenon known as eutrophication (see Section 4.5.13 below).
- In regions with significant areas of peat soils, high levels of NOM, dominated by humic material, can leach into raw water and cause particularly acute water discolouration problems and high potential for disinfection byproduct formation (see Section 4.5.14 below).

Mitigation measures can be difficult. Where natural levels of turbidity are increased due to mining, construction or other industrial activity, then this can be controlled though a requirement to manage site runoff by passage through settlement lagoons or equivalent (see later). Loss of soil from agricultural land and forestry can be reduced by improved soil and land management practice. Mitigation of hazards from eutrophication and humic materials is dealt with in later sections.

4.5.3 Natural groundwater hazards

Reactions of rainwater in the soil/rock profile during infiltration and percolation provide groundwater with its essential mineral composition. It takes up carbon dioxide, and the resultant weak acid dissolves soluble minerals. In humid climates with regular recharge, groundwater moves continuously and contact times can be relatively short, with only the most readily soluble minerals being dissolved.

Nine major chemical constituents (sodium, calcium, magnesium, potassium, bicarbonate, chloride, sulphate, nitrate and silicon) make up 99% of the solute content of natural groundwaters (Freeze & Cherry 1979). The proportion of each of these constituents, and of the associated trace elements, reflects the prevailing sub-surface groundwater flow paths and hydro-geochemical evolution of the groundwater concerned. Aquifer rock type is also important, since, for example, groundwater movement in crystalline rocks occurs relatively rapidly via joints and fractures, and the rocks themselves are generally not very soluble.

Groundwater in the recharge areas of humid regions is likely to be low in overall minerals, compared to that in arid or semi-arid regions where the combination of evaporative concentration and slower groundwater movement can produce much higher concentrations. Elevated concentrations of specific solutes can occur in certain hydrogeological settings, such as high sulphate concentrations associated with the weathering of some basement rocks or dissolution of gypsum in sedimentary sequences, hardness associated with carbonate rocks, or from association with some types of geothermal activity.

Although trace elements make up only 1% of naturally-occurring dissolved constituents in groundwater, they can sometimes make it unfit or unacceptable for consumption. At the same time many trace elements are essential for human and/or animal health in small quantities and may in part be derived from drinking water. However, the desirable concentration range may be small, and some trace elements are harmful at higher levels (e.g. fluoride). Others are potentially harmful to health, even at very low concentrations (e.g. arsenic and uranium).

Certain elements, for example arsenic and fluoride, have been identified by WHO as presenting known health risks in groundwater. Other elements are of increasing concern and may be subject to further investigation. The concentration of some of these constituents can also be increased through the impact of polluting activities at the land surface. It is, therefore, important for management purposes to differentiate anthropogenic impacts from naturally occurring problems. This will require investment in detailed groundwater investigation and monitoring. As well as those minerals of significance for health, a number will also impact on consumer acceptability, particularly those such as iron and manganese which can lead to water discolouration. Highly mineralised sources can also be unacceptable to consumers due to adverse taste or formation of hardness scale in hot drinks and hot water plumbing systems.

All groundwater sources should be analysed for mineral composition before being developed for water supply purposes. The level of key minerals should be assessed against national health guidelines and/or advice from WHO. Consumer acceptability issues may also need to be taken into account. Where levels are of potential concern, the action to be taken will depend on local circumstances, including the scale of any potential problem, the level of financial resources available, and availability of suitable alternative public supply sources. Mitigation measures fall into three main areas. If none of the options are readily available then the action taken will require more detailed evaluation in conjunction with the relevant health and other authorities:

- *Closure of source*
  In larger utilities where alternative water resources are available it might be most cost effective to simply close the source and replace it with water from other supplies.
- *Removal through treatment*
  Depending on the mineral involved and its concentration, suitable treatment may be installed.
- *Blending with other sources before supply*
  The source may be blended before distribution with water from another source with lower levels of the relevant mineral. In such situations, however, a full risk assessment should be taken to ensure that concentrations in supply can be maintained at appropriate levels, allowing for natural variation in raw water quality and the reliability of the blending arrangements.

4.5.4 Microbiological hazards

All surface waters are likely to have a wide range of microbiological hazards present. They include bacteria, viruses, protozoa and in some situations other organisms such as helminths. They originate mainly from human sewage disposal and from farm animals and their wastes (Kay et al. 2008). In some situations, wild animals can also be a significant source of microbial pollution (McDonald et al. 2008). The microbial load in raw waters can be considerably increased during storm or other adverse weather conditions (Tyrrel and Quinton 2003) and this needs to be taken into account as part of the WSP risk assessment. Much more information on the risks associated with specific organisms is available from WHO (WHO 2011).

The unsaturated and saturated zones of aquifers are usually efficient in degrading microbiological hazards. Therefore, for deep well protected aquifers, microbiological contamination in groundwater is much less common. However, some aquifers are much more vulnerable to microbial contamination e.g. those which are unconfined and with shallow water table. Additionally, inadequate design, construction or maintenance of the well and borehole headworks can allow direct ingress of microbial pollution from surface water or shallow aquifer layers.

Microbial pollution of raw water can be a major hazard for drinking water supplies. Indeed, many of the most serious incidents of waterborne disease, even in wealthy developed countries, have been due to an unrecognised increase in the microbial load in raw water, coupled with inadequate disinfection barriers (Hrudey and Hrudey 2004). Typically, a major risk can occur when one or more of the following situations arises:

- The microbial load in the raw water increases above the level that can be removed by the existing natural or disinfection barriers.
- The type of microbial load changes, e.g. increased levels of *Cryptosporidium* which the plant was not designed to cope with.
- The effectiveness of the existing disinfection barriers reduces, e.g. due to failure or poor maintenance of treatment plant, treatment processes or units being out of commission for maintenance, or an unrecognised increase in the raw water turbidity or oxidation demand which reduces disinfection effectiveness.

Because of the importance for protection of public health, the catchment WSP should ensure that there is an ongoing assessment of the risk from microbial contamination of the raw water under all types of weather conditions. Particular attention should be given to *Cryptosporidium* or other parasites because, unlike most bacteria and viruses, they are not removed or inactivated by conventional chemical disinfection. In conjunction with the catchment management authorities, new developments which might pose a risk of increased raw water microbial contamination (e.g. new human settlements, increased livestock density, or animal processing plants) should be assessed and mitigation action taken.

Except where there is absolute certainty that raw water microbial contamination cannot occur, then robust disinfection treatment processes will be necessary to ensure that the treated water remains safe at all times. If C*ryptosporidium* or other protozoa may be present, then treatment will need to ensure that the organisms are physically removed (e.g. through filtration) or inactivated. In larger, particularly surface, catchments, complete mitigation of microbial risks can be difficult and, thus, disinfection needs to be correspondingly more rigorous. However, in smaller or less developed catchments, where treatment plant may have limited disinfection, some mitigation of raw water risk may be possible, for example:

- For vulnerable groundwaters/spring sources, by fencing to restrict animal access around the immediate source catchment area, and by ensuring that well head design and construction is adequate and well maintained (see Chapter 5.3.5).
- In upland areas livestock should, if possible, be kept away from catchment feeder streams by fencing and provision of separate watering facilities.

- Where possible, temporarily ceasing abstraction or using alternative supply sources if raw water quality deteriorates to a level where disinfection may be compromised.

4.5.5 Sewage and wastewater reuse

As described above, human sewage can pose a major microbial risk to water treatment works. However, depending on the level and effectiveness of treatment, sewage disposal and sludge disposal can be a significant source of ammonia, nitrite, nitrate and natural organic material and also, in industrial catchments, a wide range of organic and inorganic pollutants. Risks can be mitigated through legally enforceable requirements to treat sewage to defined standards, manage sewer system surcharge in storm conditions, and properly treat and dispose of wastewater sludges.

In groundwater, water reuse is being increasingly adopted for managed aquifer recharge (MAR) (Dillon et al. 2010). The main objectives are to store water for future use, implement recharge rates in overexploited areas, use the aquifer to attenuate some pollutants and minimise surface runoff. There are many techniques to infiltrate sewage and wastewater, divided in two main groups: infiltration in the riverbed and outside of the riverbed. Careful evaluation of the wastewater quality is necessary in order to only inject water with contaminants that are degradable in the aquifer. Due to the aquifer contamination risks involved, MAR should only be implemented in situations where it is really necessary, such as semi-arid and arid regions, and in costal areas to control saline intrusion (Koussis et al. 2010, IAH 2011).

4.5.6 Pesticides

Use of pesticides can pose risks to raw water used to produce drinking water. The treated water standards which apply to pesticides will vary from country to country. In many places standards are based on the toxicity of individual substances (WHO 2011), but in Europe legislation sets an extremely stringent limit for pesticides (0.1 $\mu g\ L^{-1}$ for individual pesticides and 0.5 $\mu g\ L^{-1}$ for the sum of all pesticides detected). This is effectively a surrogate for zero, based on the principle that the presence of all such substances in drinking water is undesirable.

Pesticides are used particularly in arable agriculture, animal husbandry and forestry, but also for amenity purposes, e.g. control of weeds on roads and other impermeable paved surfaces, on railways, and in parks, sports facilities and gardens (Ramwell 2005, Ramwell et al. 2002 and 2004).

Pesticide losses will vary with crop and cultivation practice, chemical properties (particularly soil half life and the organic carbon to water partition coefficient), soil characteristics (particularly texture and soil organic matter content), soil condition (e.g. water content), the presence of artificial drainage and weather conditions (especially rainfall patterns in periods following application). However, except where there is bad practice, only a small proportion of agrochemicals applied are normally leached to water. Losses rarely reach 5% of total active ingredient applied and are normally less than 1% for chemicals which are weakly or moderately mobile (e.g. Foster and Hirata 1988, Johnson et al. 1996).

However, even with this potentially low leaching loss, pesticides can still pose significant risks for drinking water, particularly in those areas which have very strict standards, such as Europe. The risks are typically higher in surface water, particularly from the use of mobile and persistent products in agriculture. Pesticides from agriculture can reach surface water via a number of routes:

- Overspray of streams and ditches or wind driven spray drift to water courses.
- Poor storage and handling in farmyards resulting in spillage and subsequent wash off.
- Losses in overland flow during intensive rainfall or from saturated or frozen ground.

- Leaching through soil horizons to shallow or deep groundwater and subsequent transport to surface water bodies.
- Loss through field drainage where installed.

Pesticides can also pose a risk to groundwater, particularly where the potential attenuation effect of soil is bypassed, for example through poor storage or disposal practice, or excessive or incorrect application on drained hard surfaces such as roads, railways or other vulnerable locations.

Effective mitigation of pesticide risks first requires a good knowledge of pesticide usage in the catchment, including which active ingredients are applied, the application rate, for what purpose, when and how. This must be associated with a knowledge of pesticide properties, mapping of the agricultural and other land use in the catchment, as well as the relevant soil properties and underlying geology. Finally, the hydrological response of the catchment to precipitation events needs to be described. This information is used in several models (Brown et al. 2002, Kannan et al. 2006) which are able to predict land to water transfers of pesticides and associated expected exposure (concentration and duration) in surface or groundwater sources. There is an increasing amount of best practice information available on how losses from all sources can be reduced (VI 2011), but this requires the active cooperation of users and other catchment stakeholders. In many countries there are also rigorous legislative requirements for pesticides which may require specific authorisation for use and storage of pesticides, as well as conditions for use, including rate of application, timing of application, or restrictions for some or all purposes. There may also be statutory requirements for training/certification of users and disposal of used containers.

4.5.7 Nitrate

The WHO guideline value for nitrate is 50 mg/l as NO_3. In many parts of the world this value is regularly exceeded, or is at risk of being exceeded, in raw drinking water resources, particularly groundwater. The causes can be complex but mainly originate from intensive farming, including intensive grassland, which relies on high inputs of manures and inorganic fertilisers and which can enhance nitrogen mineralisation and nitrate formation by cultivation practices such as ploughing (Whitmore et al. 1992, Whelan et al.1995, Howden et al. 2011). Ploughing of permanent grassland makes a particularly important contribution to nitrate leaching. Equilibrium levels of soil organic matter under permanent grassland are much higher than under arable land on similar soil types, and much of this carbon and nitrogen is mineralised by ploughing. Whitmore et al. (1992) estimated that soil organic nitrogen concentrations in UK soils decline by approximately 50% in the first 5.5 years after ploughing, and by 90% after 18 years. In total up to 4 t N ha^{-1} can be released from the top 25 cm in the first 25 years after ploughing, and much of this N will be vulnerable to leaching (Whitmore et al. 1992). Nitrate contamination can also arise from poorly controlled treatment and disposal facilities for animal wastes, sewage and sewage sludge.

Extensive areas of arable monoculture can generate the most serious diffuse contamination of groundwater by nitrate. More traditional crop rotations and extensive pasture land normally present lower risks of a sub-surface contamination with nitrate. Leaching losses from organic agriculture are often perceived to be low but can sometimes be as high as from conventional systems, especially when expressed per unit of agricultural product (Mondelaers et al. 2009). Agriculture involving the cultivation of perennial crops is normally associated with lower leaching losses than where seasonal cropping is practiced, because there is less disturbance and aeration of the soil and a more continuous plant cover, over the winter season when leaching is most intense, with associated demand for nutrients. However, when perennial crops have to be renewed and the soil ploughed, there can be a major release and leaching of nutrients (N and P). Values of leaching losses obtained from the literature (Foster & Hirata 1988) indicate that up to 75% of the total N applied can be transformed to nitrate and leached to groundwater (although values of 50% are more commonly cited). In some areas other potential sources of groundwater nitrate contamination need to be considered. For example, in arid and semi-arid climates, agricultural irrigation with wastewater can be an issue. Wastewaters invariably contain nutrients and salts in excess of crop requirements, which can lead to significant leaching losses from agricultural soils, as well as causing salinity problems for the soil itself.

As with pesticides, mitigation of nitrate losses requires a good knowledge of land use in the catchment, as well as mapping soil types and underlying geology. A wide range of models have been developed which enable nitrate concentrations in surface or groundwater to be predicted (Whelan et al. 1995, Smith et al. 1996, Lord and Anthony 2000). Mitigation measures are normally based on controlling cropping, fertiliser use and timing (especially with respect to crop demands for nitrogen) and land use changes to reduce leaching losses, but this often requires legislation to be effective (Hiscock et al. 2007, Cherry et al. 2008).

In urban areas, significant nitrate pollution of groundwater can derive from leakage of existing sewer systems (Wakida and Lerner 2005) and this can be a particular problem where sewerage infrastructure is poor or non existent. In such situations, local sanitation arrangements (latrines and septic tanks) can present significant aquifer quality hazards which need to be managed, especially in high population density areas. In most aquifer types, except the vulnerable shallow sources, there should be sufficient natural groundwater protection to eliminate faecal pathogens in percolating wastewater from properly constructed *in situ* sanitation. However, elevated concentrations of nitrogen compounds (usually nitrate) will also be present at various levels according to the population density served by latrines and septic systems. Mitigation requires long term investment to install mains sewerage or, in specific areas, to improve septic tank systems and field drains, if present.

4.5.8 Urban development and runoff

As described above, urban development, particularly in less affluent countries, can be a significant source of groundwater nitrate. Additionally, runoff of rainwater from roads in urban areas can lead to the transfer of a complex mixture of pollutants containing phosphorus, nitrate, ammonia, heavy metals, pesticides, hydrocarbons, oil and greases, and other inorganic and organic material. These pollutants come from a variety of sources, including atmospheric deposition, motor vehicle corrosion, tyre wear and combustion products (Napier et al. 2008), chemical spills, and animal faeces (O'Keefe et al. 2005). Concentrations of many urban contaminants have been shown to be particularly high during the so-called "*first flush*" (Lee et al. 2007), when intense storms follow a very dry period over which contaminants accumulate. Water suppliers need to assess the risks associated with these pollutants as part their catchment WSP. Mitigation can be difficult and expensive but, where the risks are significant, could involve for example, provision of oil traps or interception and treatment of "*first flush*" storm water.

4.5.9 Fuel storage and transport

Fuel storage and transport, either though pipelines or by road, represent an important raw water contamination hazard for both surface waters and groundwaters used for drinking water production. When oil leaks or is spilled, the often complex mixture of hydrocarbons can present a significant toxic hazard to waterworks abstractions. Even at low concentration, hydrocarbons can often lead to taste complaints in treated water, in addition to contributing to potential health risks to consumers. If the soil and underlying groundwater is severely contaminated with oil residues, this can have an impact for decades. The problem is often related to corrosion leading to leakage from storage tanks and pipe work. This can be due to a number of factors, including poor design, construction and maintenance.

Accidental oil tanker spills can occur in routine operations of fuel loading, unloading and provisioning. Other diffuse sources, e.g. from highway drainage systems, must also be considered (see above).

Mitigation of risks from oil leaks and spills can be secured through a range of either voluntary or statutory best practice measures, underpinned by risk based procedures to identify and correct potentially corroded fuel storage and transport infrastructure before it leaks. This can be helped by a risk classification of different plant and equipment linked to regular inspection and maintenance procedures. For example, tank corrosion is often associated with age and there is a strong correlation that tanks more than 20 years old are often seriously corroded and subject to leaks, unless they receive regular maintenance. Additional security can be obtained by design of oil storage facilities so that even if leaks occur, the oil can be contained on site without causing pollution. For example, tanks installed above ground should have leak proof bunds sufficient to contain at least the volume of oil stored. Level indicators and alarms can warn of potential problems. If tanks must be kept underground, then they should be double skinned and, again, be fitted with level indicators and alarms. Drainage from sites with oil storage and/or subject to regular vehicle use should have oil traps to protect surface water drainage systems.

In certain countries, large rivers and canals are used extensively for commercial shipping with the consequent risk of oil leakage. This can be mitigated by voluntary or legal requirements to better manage fuel/oil storage on board, and also by restrictions on discharge of polluted wastewater, other than at authorised locations (e.g. at docks) with proper disposal treatment facilities.

Smaller-scale disposal of waste mineral oil (e.g. from motor vehicle oil changes) to surface water drains or sewers can also result in water pollution incidents. This can be managed via awareness raising campaigns and recycling schemes (D'Arcy et al. 1998).

4.5.10 Industry

The range of industrial contaminants that can affect water supply abstractions is potentially very large. As well as potential toxic threats, many industrial substances can also present issues for water suppliers because of their potential to cause unacceptable tastes and odours at levels well below that which might give rise to a toxic effect. For example, it was found that even a few hundred litres of some solvents or hydrocarbons spilled many kms upstream of a surface water intake, gave rise to tastes in final drinking water (Gray and Thompson 2004). This can also occur with substances such as phenols, which can react with chlorine to create unacceptable tastes in the final water distributed to consumers. Some substances, such as chlorinated or other solvents, can pose a significant risk to groundwater if they enter sub-surface layers.

Because of the range and complexity of industrial installations, particularly in large catchments, risk assessment and mitigation can be difficult and time consuming. However, a typical approach could include creation, by the catchment management authorities, of an inventory of all significant industrial installations and activities within the catchment. Those with particular risks to water supply abstractions and/or the environment generally should then be subject to statutory or voluntary controls to reduce the risk that chemical or other hazards could enter the aquatic or sub-surface environment at levels of concern. The control methods used will vary, but could include one or more of the following:

- Treatment of process and other harmful wastes, with discharge controlled through permits specifying volume and concentration of pollutants.
- Permits specifying requirements for biological wastes discharged to sewer for treatment at a sewage works, e.g. maximum metal limits.

- Requirements covering storage and disposal of hazardous material which might pose a risk to surface water, soils or groundwater.
- Requirements that surface water drainage from industrial sites be subject to treatment through interceptors to contain any spillage on site.
- Measures to minimise the risk of leaks from chemical storage tanks and pipelines though risk based inspection and maintenance, coupled with bunding requirements to contain spills that do occur.
- Reformulation, or restricting the use of, or even banning certain particularly harmful and persistent chemicals. This could include those known as Persistent Organic Pollutants (POPs) used in industry and agriculture. It could also cover products such as pharmaceuticals (WHO 2011b) and cleaning products which might enter surface water through sewer systems and present a potential hazard to water supply intakes.

4.5.11 Solid waste disposal

Inadequate disposal of solid waste is responsible for a significant number of cases of groundwater pollution (USEPA 2011). In some situations, leachate from waste sites may also pose a risk to surface water abstractions. This problem is more prevalent in wet regions where substantial volumes of leachate may be generated from these facilities, but it also occurs in more arid climates where leachates will generally be more concentrated. The sub-surface contaminant load generated from a waste tip or sanitary landfill is a function of two factors:

- The probability of the existence of contaminants in the solid waste.
- The generation of a hydraulic gradient sufficient to leach such contaminants.

Measures to mitigate this potential source of contamination relate mainly to ensuring that both the base and surface of the solid waste disposal area are impermeable. This should be a requirement for any newly constructed sites, but it may not be practical for old pre-existing sites. In such situations, and for all new sites, this will normally also require the collection and treatment of leachates. It is also good practice to monitor groundwater in the vicinity of waste sites to assess any potential risks to aquifers.

4.5.12 Mining and mineral extraction

The activities of extractive industry include a number of operations that can provide a water quality risk to both surface and groundwater potable abstractions (Thompson et al. 2007). These include:

- Exploration and development for mineral and petroleum resources often involves land clearing which may lead to contaminated stormwater runoff. Risks may also occur from spills and leaks from fuel and waste disposal.
- Oxidation and leaching of minerals from mining spoil and other waste products can contaminate groundwater and surface water.
- Processing of minerals using a variety of mechanical and chemical treatment processes can result in a risk of water contamination from storage ponds holding processing liquors, and tailings dams.
- Acid mine drainage (AMD) is probably the most severe environmental problem that can occur on mine sites. It happens where mineral and coal deposits containing sulphide minerals are exposed to air and oxidised, releasing sulphuric acid.

Mitigation options usually require properly designed storage and treatment facilities to effectively manage effluent and site runoff to standards necessary to prevent water pollution. On site management procedures must include dealing with emergencies and plant breakdown.

4.5.13 Algae and algal toxins

Proliferation of algae due to eutrophication (nutrient enrichment) of surface waters, especially in lakes and reservoirs, can lead to tastes/odours which are unacceptable to consumers. They can also lead to increased oxidation demand and, thus, impact on disinfection efficiency and disinfection byproduct formation. Of particular importance is the fact that the presence of some algae has also been recognised as a serious potential human health risk. High concentrations or blooms of Cyanobacteria (blue-green algae), such as *Microcystis* and *Anabaena*, can occur in slow-moving or still surface waters with a moderate to high concentration of nutrients, particularly phosphorus. In perhaps 50% of cases these can produce a range of toxins (cyanotoxins) which, if ingested through drinking water, can represent a risk to human health (Chorus and Bartram, 1999).

Mitigation measures for cyanobacterial bloom formation need to address the growth requirements for the organisms, by control and reduction of external nutrient loading to the water body, and where possible light availability (e.g. artificial mixing). The most significant sources of external nutrient inputs are often runoff and erosion from fertilised agricultural areas, erosion resulting from deforestation, and sewage (treated and untreated). Mitigation may, therefore, be achieved via improvement of agricultural land use and practices, and by applying advanced wastewater treatment methods for nutrient removal. The latter are most likely to result in immediate and tangible benefits (Parker and Maberly 2000).

Management of raw water abstraction can also be effective in reducing the amount of cyanobacterial cells entering the water supply system. This can be achieved by choosing an optimum position (site and depth) for pumped abstractions from surface water bodies, or where practical, by abstracting surface water through bank filtration. In addition, it has been shown that the use of submerged barley straw can help to mitigate adverse impact of algal blooms, if properly installed around surface abstraction points (Everall and Lees 1996 and 1997, Garbett 2005, Murray et al. 2010). As the straw decomposes, algistatic compounds are released which limit the development of the bloom (Murray et al. 2010).

4.5.14 Colour in upland catchments

The colour of raw surface water is usually classified as either "*real*" colour caused by dissolved matter or "*apparent*" colour due to the presence of suspended or colloidal material. The most common colours of natural waters are yellow and brown and are caused by the presence of NOM, particularly compounds known as humic substances originating from organic soils, such as peats, and decayed vegetation. Four groups of humic substances can be distinguished based on their solubility in various solvents: fulvic acids (about 90% of total humic substances), hymatomelanic acids, humic acids and humus coal. In some waters, the brown colour is enhanced by the presence of iron and manganese.

Raw water colour can typically be removed by conventional coagulation, clarification and filtration. However when chemical oxidation e.g. by chlorine or ozone is used then care has to be taken to avoid excessive levels of disinfection byproducts, such as trihalomethanes (Lewin et al. 2004).

In some peat moorland catchments the level of colour in runoff can be significantly increased by human factors including burning of vegetation, and drainage, as well as weather effects such as drought and changes in atmospheric deposition. Peat soils form originally under wet conditions, but when they dry physicochemical changes may occur to the organic matter which affect its ability to adsorb water, and which may affect the release of humic substances. Severe drying of peat can enhance microbially mediated decomposition, which can be leached by rainfall when water tables rise in the post drought period (Mitchell and McDonald, 1992). Some upland areas have been drained artificially and this has also been linked to enhanced water discolouration. Recently, drain-blocking has been encouraged in many UK upland areas to promote peatland habitats and a number of studies have been performed to test the effects of this on DOC (dissolved organic carbon) production. Wallage et al. (2006) observed higher DOC losses from drained slopes compared with from intact peat and drain-blocked slopes, although this may not always occur (Worrall et al. 2007). Burning of vegetation may also affect peat moisture content, physical structure and microbial activity (Yallop and Clutterbuck 2009) but this may not always be the case (Clay et al. 2009). Monteith et al. (2007) have hypothesised that recent increases in DOC concentrations may also be a response to a decrease in the sulphate content of atmospheric deposition.

In many cases, it is therefore likely that water colour is influenced by several complex factors acting together and may be confounded by site-specific and or regional scale factors. Thus mitigation measures will have to be determined locally.

4.5.15 Radiological

Health risks from radioactive contaminants in drinking water abstracted from surface water or groundwater sources can arise from a range of industrial, medical and military related activity in the contributing catchment. Significant risks will generally only be associated with the presence of sites using nuclear material, although there may be some contribution of natural radioactive isotopes in some catchments, for example those containing volcanoes (Kozlowska et al. 2009). Mitigation of risks is similar to that for other industrial contaminants, although the hazardous nature of the material means that controls over use and emissions are already likely to be stringent.

However, due to its percolation through the soil, groundwater sources of radioactive contaminants may be more vulnerable than surface waters compounds due to the dissolution of elements such as uranium and other naturally-occurring radiological elements. The actual risk can only be assessed by evaluation of the naturally occurring elements (see Section 4.5.3 above).

4.5.16 Saline intrusion

Excessive abstraction can result in the intrusion of saline water and sometimes other polluted groundwater into a productive freshwater aquifer because of a reduction or reversal in hydraulic gradients. This situation is common in coastal aquifers (marine intrusion), but can also be problematic inland with contaminated aquifers. The main contaminant of concern is chloride, but persistent anthropogenic contaminants, including chlorinated solvents, may also be transported into the aquifer via this mechanism. The risk of such intrusions can be reduced by use of flow models which allows a calculation of the extraction rate that will not induce contaminated water into the aquifer.

4.5.17 Recreation

Some raw water storage reservoirs used for drinking water production are in protected catchments where human access is strictly controlled. In other situations drinking water reservoirs may be used for recreational purposes, for example boating, sailing or fishing. Depending on the extent of such recreational activity and the level and type of treatment available at the water treatment plant, this activity could, in some situations, pose some risk to raw water quality. Any potential risk needs to be assessed and appropriate mitigation measures adopted. For example, restricting the use of petrol powered boats can reduce the risk of fuel leakage.

REFERENCES

Arnscheidt, J., Jordan, P., Li, S., McCormick, S., McFaul, R., McGrogan, H. J., Neal, M., Sims, J. T. (2007) Defining the sources of low-flow phosphorus transfers in complex catchments. *Science of the Total Environment* **382**, 1–13.

Bartram, J., Corrales, L., Davison, A., Deere, D., Drury, D., Gordon, B., Howard, G., Reinehold, A., Stevens, M. (2009) Water Safety Plan Manual. Step-by-Step Risk Management for Drinking-water Suppliers. World Health Organisation, Geneva, Switzerland.

Brown, C.D., Bellamy, P.H. and Dubus, I.G. (2002) Prediction of pesticide concentrations found in rivers in the UK. *Pest Management Science* **58,** 363–373.

Burt T. P., Howden, N. J. K., Worrall F. and Whelan M. J. (2008) Importance of long-term monitoring for detecting environmental change: Lessons from a lowland river in south east England. *Biogeosciences* **5**, 1529–1535.

Burt T.P., Howden, N.J.K., Worrall F. and Whelan M.J. (2010) Long-term monitoring of river water nitrate: how much data do we need? *Journal of Environmental Monitoring* **12**, 71–79.

Cherry, K.A., Shepherd, M., Withers, P. J. A. and Mooney, S. J. (2008) Assessing the effectiveness of actions to mitigate nutrient loss from agriculture: A review of methods. *Science of the Total Environment* **406**, 1–23.

Chorus, I. and Bartram J., (1999) Toxic Cyanobacteria in Water: A Guide to Their Public Health Consequences, Monitoring and Management. World Health Organisation, Geneva, Switzerland.

CCME (2004) From Source to Tap: Guidance on the Multi-Barrier Approach to Safe Drinking Water. Canadian Council of Ministers of the Environment, Winnipeg, Manitoba, Canada.

Clay C.D., Worrall F. and Fraser E.D.G. (2009) Effects of managed burning upon dissolved organic carbon (DOC) in soil water and runoff water following a managed burn of a UK blanket bog. *Journal of Hydrology* **367**, 41–51.

D'Arcy B.J., Usman, F., Griffiths, D. and Chatfield, P. (1998) Initiatives to tackle diffuse pollution in the UK. *Water Science and Technology* **38**, 131–138.

Dillon, P., Toze, S., Page, D., Vanderzalm, J., Bekele, E., Sidhu, J., Rinck-Pfeiffer, S. (2010) Managed aquifer recharge: rediscovering nature as a leading edge technology. *Water Science and Technology* **62**, 2338–2345.

Everall, N.C. and Lees, D.R. (1996) The use of barley straw to control general and blue-green algal growth in a Derbyshire reservoir. *Water Research*, **30**(2), 269–276.

Everall, N.C. and Lees, D.R. (1997) The identification and significance of chemicals released from decomposing barley straw during reservoir algal control. *Water Research*, **31**(3), 614–620.

Foster, S. S. D. and Hirata, R., (1988) Groundwater pollution risk assessment: a methodology using available data. WHO-PAHO/HPECEPIS Technical Manual. Lima, Peru.

Foster, S. S. D., Adams, B., Morales, M., and Tenjo, S., (1993) Groundwater protection strategies: a guide towards implementation. UK ODA, CPR, WHO/PAHO-HPE Technical Manual. Lima, Peru. p88.

Foster, S., Hirata, R., Gomes, D., D`Elia, M., Paris, M., (2002) Groundwater Quality Protection, The World Bank, Washington DC, p103.

Freeze, A. and Cherry, J. (1979) Groundwater. Published by Prentice-Hall, Englewood Cliffs.

Garbett R. (2005) An investigation into the application of floating reed bed and barley straw techniques for the remediation of eutrophic waters. *Water and Environment Journal* **19**, 174–180.

Gray G. and Thompson K. C. (2004) Water Contamination Emergencies - Can we cope? Royal Society of Chemistry, London

Haria, A.H., Johnson A.C., Bell, J.P. and Batchelor, C.H. (1994) Water-Movement and Isoproturon Behaviour in a Drained Heavy Clay Soil .1. Preferential Flow Processes *Journal of Hydrology* **163**, 203–216.

Hiscock, K., Lovett, A., Saich, A., Dockerty, T., Johnson, P., Sandhu C., Sünnenberg, G., Appleton K., Harris, B., and Greaves, J. (2007) Modelling land-use scenarios to reduce groundwater nitrate pollution: the European *Water4All* project, *Quarterly Journal of Engineering Geology & Hydrogeology*, **40**(4), 417–434.

Holman, I.P. Palmer, R.C., Bellamy, P.H., Hollis, J.M. (2005) Validation of an intrinsic groundwater pollution vulnerability methodology using a national nitrate database. *Hydrogeology Journal* **13**, 665–674.

Howden N.J.K., Burt T.P., Mathias S.A., Worrall F., Whelan M.J. and Bieroza M. (2011) Modelling long-term diffuse nitrate pollution at the catchment-scale: data, parameter and epistemic uncertainty, *Journal of Hydrology* in press.

Hrudey, S. E. and Hrudey, E. J. (2004), Safe Drinking Water: Lessons from Recent Outbreaks in Affluent Nations, IWA Publishing, London.

IAH International Association of Hydrogeologists 2011, IAH-MAR Managed Aquifer Recharge Web forum, http://www.iah.org/recharge

Johansson, P.O. and Hirata, R., (2001) Rating of groundwater contamination sources. In: Zaporozec, A., Groundwater contamination inventory: A methodological guideline. Paris, France: UNESCO.

Johnson, A. C., Haria, A. H., Bhardwaj, C. L., Williams, R. J. & Walker, A. (1996) Preferential flow pathways and their capacity to transport isoproturon in a structured clay soil. *Pesticide Science*, **48** (2) 25–237.

Kannan N., White S.M., Worrall F. and Whelan M.J. (2006) Pesticide modelling for a small catchment using SWAT-2000, *Journal of Environmental Science and Health Part B* **41**, 1049–1070.

Kay D., Crowther, J., Fewtrell, L. Francis, C.A., Hopkins, M., Kay, C., McDonald, A. T. Stapleton, C. M., Watkins, J., Wilkinson, J. and Wyer, M. D. (2008) Quantification and control of microbial pollution from agriculture: a new policy challenge? *Environmental Science & Policy* **11**, 171–184.

Kozlowska, B., Morelli, D., Walencik, A., Dorda, J., Altamore, I., Chieffalo, V., Giammanco, S., Imme, G. and Zipper, W. (2009) Radioactivity in waters of Mt. Etna (Italy). *Radiation Measurements* **44**, 384–389.

Koussis, A. D., Georgopoulou, E., Kotronarou, A., Mazi, K., Restrepo, P., Destouni, G., Prieto, C., Rodriguez, J. J., Rodriguez-Mirasol, J., Cordero, T., Ioannou, C.. Georgiou, A., Schwartz, J., Zacharias, I. (2010) Cost-efficient management of coastal aquifers via recharge with treated wastewater and desalination of brackish groundwater: application to the Akrotiri basin and aquifer, Cyprus, *Hydrological Sciences Journal* **55**, 1234–1245.

Lee, H., Swamikannu, X., Radulescu, D., Kim, S. and Stenstrom, M.K. (2007) Design of stormwater monitoring programs *Water Research* **41**, 4186–4196.

Lewin, N., Zhang, Q., Chu, L.L. and Shariff, R. (2004) Predicting total trihalomethane formation in finished water using artificial neural networks *Journal of Environmental Engineering and Science* **3**, S35–S43.

Lord E.I. and Anthony S.G. (2000) MAGPIE: A modelling framework for evaluating nitrate losses at national and catchment scales, *Soil Use and Management* **16,** 167–174.

McDonald, A.T., Chapman, P.J. and Fukasawa K. (2008) The microbial status of natural waters in a protected wilderness area, *Journal of Environmental Management* **87**, 600–608.

Mitchell, G. and McDonald, A.T. (1992) Discoloration of water by peat following induced drought and rainfall simulation, *Water Research* **26**, 321–326.

Mondelaers, K, Aertsens, J., Van Huylenbroeck, G. (2009) A meta-analysis of the differences in environmental impacts between organic and conventional farming, *British Food Journal* **111,** 1098–1119.

Monteith et al., (2007) Dissolved organic carbon trends resulting from changes in atmospheric deposition chemistry. *Nature*, **450**, 537.

Murray, D., Jefferson, B., Jarvis, P. and Parsons, S.A. (2010) Inhibition of three algae species using chemicals released from barley straw. *Environmental Technology* **31,** 455–466.

Napier, F., D'Arcy, B. and Jefferies, C. (2008) A review of vehicle related metals and polycyclic aromatic hydrocarbons in the UK environment. *Desalination* **226,** 143–150.

O'Keefe, B., D'Arcy B.J., Davidson, J., Barbarito, B. and Clelland, B. (2005) Urban diffuse sources of faecal indicators. *Water Science and Technology* **51,** 183–190.

Parker, J.E. and Maberly, S.C. (2000) Biological response to lake remediation by phosphate stripping: control of Cladophora. *Freshwater Biology* **44,** 303–309.

Ramwell, C.T. (2005) Herbicide sorption to concrete and asphalt. *Pest Manag. Sci.* **61**, 144–150.

Ramwell, C.T., Heather A.I.J. and Shepherd, A.J. (2002) Herbicide loss following application to a roadside. *Pest Manag. Sci.* **58**, 695–701.

Ramwell, C.T., Heather A.I.J. and Shepherd, A.J. (2004) Herbicide loss following application to a railway. *Pest Manag. Sci.* **60**, 556–564.

Smith J.U., Bradbury N.J., and Addiscott, T.M. (1996) SUNDIAL: A PC-based system for simulating nitrogen dynamics in arable land. *Agronomy Journal* **88,** 38–43.

Stollard, R.J. and Rickard, A. (2005) West country Rivers Trust United Kingdom: A pioneering programme for restoration and regeneration of major river basins in Omelchenko A; Pivovarov AA; Swindall WJ (eds.)

Modern Tools and Methods of Water Treatment for Improving Living Standards pp 265–278, NATO Science Series IV Earth and Environmental Sciences.

Thompson, T., Fawell J., Kunikane, S., Jackson, D., Appleyard, S., Callan, P., Bartram, J. and Kingston, P. (2007) Chemical safety of drinking-water: Assessing priorities for risk management. Published by WHO, Geneva.

Tyrrel, S.F., and Quinton, J.N. (2003) Overland flow transport of pathogens from agricultural land receiving faecal wastes. *Journal of Applied Microbiology* **94**, 87S–93S.

UK Environment Agency (1999) Methodology for the derivation of remedial targets for soil and groundwater to protect water resources. Access via *http://www.environment-agency.gov.uk/research/planning/40373.aspx*

USEPA, (1994) Handbook on groundwater and wellhead protection. Report W/1994. Washington, D.C.

USEPA, (2008) Handbook for Developing Watershed Plans to Restore and Protect Our Waters. EPA 841-B-08-002. United States Environmental Protection Agency. Washington, DC.

USEPA (2011) Waste facilities website http://www.epa.gov/epawaste/index.htm

VI-The UK pesticides voluntary initiative (2011) Access via http://www.voluntaryinitiative.org.uk/

Vieira, J.M.P., and Morais, C. M. (2005) Planos de Segurança da Água para Consumo Humano (Drinking Water Safety Plans). IRAR Edition. Lisbon, Portugal.

Vieira, J.M.P, Pinho, J.L.S., and Lima, M.M.L., (2006) ODEANA – Um Ambiente Hidroinformático de Suporte à Decisão na Gestão da Água numa Bacia Hidrográfica (*A Hydroinformatic Platform for Decision Support Systems in a River Basin*). In Proceedings of Portuguese Water Congress, 13–17 March, Figueira da Foz (Portugal)

Wakida, F.T. and Lerner, D.N. (2005) Non agricultural sources of groundwater nitrate: a review and a case study. *Water Research* **39**: 3–16.

Wallage Z. McDondald A.T. and Holden J. (2006) Drain blocking: An effective treatment for reducing dissolved organic carbon loss and water discolouration in a drained peatland. *Science of the Total Environment* **367**, 811–821.

Whelan M.J., Kirkby M.J. and Burt T.P. (1995) Predicting Nitrate Concentrations in Small Catchment Streams in Trudgill S.T. (ed.) *Solute Modelling in Catchment Systems* John Wiley and Sons, Chichester, pp 165–192.

Whelan, M., Hallett, S., Burt, T., Howden, N. and Worrall, F. (2011) Monitoring of diffuse pollution from agriculture – challenges and opportunities for the protection of drinking water quality, *Water and Sewerage Journal.*

Whitmore, A.P., Bradbury, N.J. and Johnson, P.A. (1992) Potential contribution of ploughed grassland to nitrate leaching. *Agriculture, Ecosystems and Environment* **39**, 221–233.

World Health Organisation (2011) Guidelines for Drinking Water Quality, 4th Edition, WHO, Geneva.

World Health Organisation (2011b) Information sheet: Pharmaceuticals in drinking-water, WHO, Geneva.

Worrall F., Armstrong, A, Holden J. (2007) Short-term impact of peat drain-blocking on water colour, dissolved organic carbon concentration, and water table depth. *Journal of Hydrology* **337**, 315–325.

Yallop, A.R. and Clutterbuck, B. (2009) Land management as a factor controlling dissolved organic carbon release from upland peat soils 1: Spatial variation in DOC productivity. *Science of the Total Environment* **608** (407), 3803–3813.

Chapter 5

Developing a treatment water safety plan

Rui Sancho and Chris Rockey

5.1 INTRODUCTION

The effective treatment of drinking water underpins the success of a water safety plan. In most situations it provides the primary barrier or series of barriers to prevent hazards that may be present in the raw source water impacting on consumers. When implementing a water safety plan for treatment processes, consideration must be given to:

- All the potential risks present in the raw water taking account of its natural variability, the pollution risks potentially present within the catchment and any risks associated with raw water storage.
- The risk of failure or reduced effectiveness of any of the treatment processes.
- The materials and chemicals used within each treatment stage.
- The risk of unintended by-products that may be produced by certain treatment processes.

This is summarised in Figure 5.1 below. Each risk needs to be assessed, mitigation measures to reduce the risk put in place, and control points established to verify that the mitigation measures are working effectively (see Chapter 3). This in turn is heavily dependent on having in place an accurate, comprehensive and well documented treatment process diagram showing all inter-linkages and critical control points.

To implement a WSP for water treatment it is necessary to clearly determine the treated water quality goals, including its acceptability to consumers (see Chapter 1). In doing so, water suppliers must also take into account the quality requirements of the final treated water in order to manage the risk of quality deterioration that might arise during onward distribution and within consumer plumbing systems (see Chapters 6 and 7). Depending on the level of risk posed by the raw water, and therefore the necessary scale and complexity of the water treatment plant and processes, a treatment WSP will require variable levels of resource and expertise to develop and maintain.

It is particularly important to assess the specific treatment challenges caused by natural variation in raw water quality and the pollution risks that exist within the catchment (see Chapter 4). Routine risk based monitoring programmes can help to build up a realistic assessment of the typical raw water quality and the extent of any variation that can be expected. The larger the data set, the higher the level of confidence in making an assessment of risk. For waters with variable quality or those that are at a high risk of pollution, consideration should be given to intensive monitoring during periods when the risk is believed to be highest. This data allows the treatment plant to be properly designed and operated with sufficient spare capacity or headroom to deal adequately with predicted worst raw water quality (see Figure 5.2 below). If suitable barriers and control measures are not present in the water treatment plant, an action plan must be developed and prioritised to reduce or eliminate potential risks.

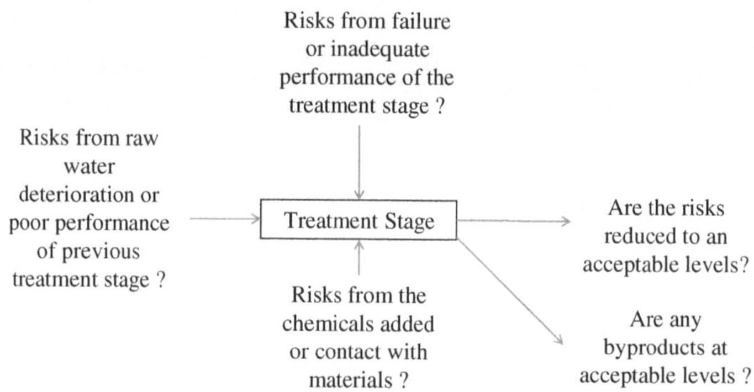

Figure 5.1 Types of risk that need to be assessed as part of a treatment WSP

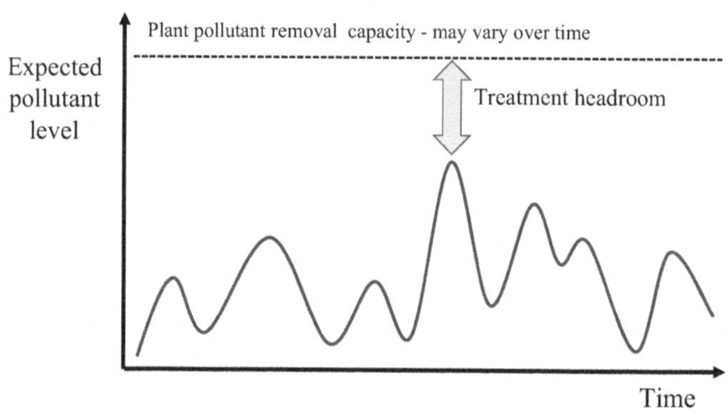

Figure 5.2 Matching expected raw water quality and treatment plant capability

Operation of water treatment processes and plant is a vitally important and skilled job. Those people operating and maintaining such plant should not only be fully trained and competent in the tasks they must perform, but also be fully aware of the water safety plan and actively contribute to its implementation (see Chapter 2). They must also be fully aware of emergency procedures and response to factors outside their normal control. All operational tasks should be fully described in Standard Operating Procedures contained in a Works Operation Manual with defined task schedules. Effective hygiene training must be in place for all staff and contractors who have the potential to affect treated water quality (see Chapter 3).

This chapter describes some of the issues that need to be considered when developing a WSP for a water treatment plant. It is not intended to cover the technology of water treatment or treatment plant design, which is dealt with in many other excellent publications. No two treatment WSPs will be the same (Figure 5.3), but the aim of this chapter is to outline the principles of developing a treatment WSP for the more common water treatment processes. It does not try to cover every one of the large number of different treatment processes that may be used in various combinations. The Chapter does describe the importance of an overall multi barrier approach and then reviews typical water quality risks, mitigation measures and control points for a range of common water treatment technologies used for both surface and groundwater treatment. Finally, the importance of plant monitoring and control are described, along with approaches to setting operational water quality targets to ensure final water quality is acceptable. Although typical approaches and solutions will be described, it is important to realise that this can only be broad guidance which needs to take into account the specific local factors and regulatory requirements for each country.

Figure 5.3 A typical water treatment plant

5.2 THE MULTI BARRIER APPROACH TO WATER TREATMENT

A water safety plan is underpinned by what is often described as a "*multi barrier approach*". This relies on having in place a number of separate sequential barriers to hazards that may be present in raw water or introduced in the water treatment plant. When designed effectively this means that if one barrier fails for any reason then the other barriers can still prevent or minimise contamination of the final water supplied to consumers. This has been defined (Health Canada 2002) as "*An integrated system of procedures, processes and tools that collectively prevent or reduce the contamination of drinking water from source to tap in order to reduce risks to public health*".

The multi barrier approach covers all aspects of water quality management, including catchment protection, water treatment, and management of the distribution network. It is supplemented by good water quality monitoring, effective emergency response systems and the pro-active education of consumers and those who work on domestic plumbing systems. It is particularly important for microbiological hazards, but can also apply to many chemical hazards.

A key element in the multi barrier approach is to ensure that the raw water within the catchment is protected and maintained at as high a quality as practicable (see Chapter 4). For some relatively small groundwater sources where source protection can be reliably secured this in itself can provide an effective barrier against most contamination. This in turn means that the water treatment applied can be relatively simple. However, for larger or more complex groundwater sources and for most surface water sources, complete catchment protection is much more difficult. In this situation there is likely to be a need for a number of integrated water treatment stages so that if deterioration or failure of a process stage occurs then the other stages can still act as effective barriers against a range of water quality hazards. See illustration in Figure 5.4.

This is an important concept as it provides the basis for comprehensive treatment of water and a system to prevent a treatment failure due to a breakdown of a single process subsequently impacting water safety. For instance, with a system which comprises coagulation/flocculation/settling, followed by rapid sand filtration and terminal disinfection, failure or deterioration in the performance of a rapid sand filter does not necessarily mean that unsafe water will be supplied. The coagulation/flocculation/settling process will remove a great deal of the suspended particles, and therefore a significant proportion of the microorganisms in the water, and the terminal disinfection will remove many of the remainder. Provided deterioration in the performance of the filtration stage is identified and rectified *quickly*, there may be little overall decrease in water quality. Similarly if the performance of an initial pre-disinfection stage (e.g. pre-filter oxidation), which would normally lower the challenge on the subsequent treatment stages, is reduced, then safe final water may be secured for a short period by relying on the final chlorine disinfection stage, perhaps with a temporary increase in chlorine dose.

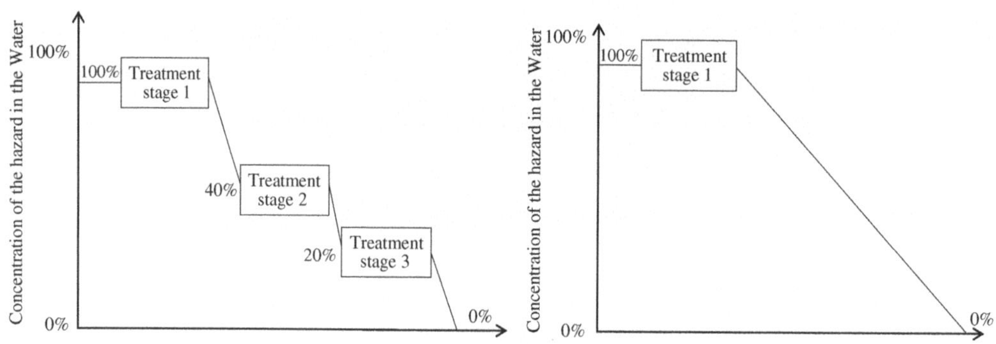

Figure 5.4 Examples of multi barrier and single barrier water treatment process

Different kinds of treatment stages can be utilised in isolation or combination to reduce different hazards to an acceptable level. It should also be recognised that the same effect can often be achieved by the use of a different sequence/combination of treatment stages. For some hazards, application of a single technology on its own may not be enough to reduce the hazard to the acceptable level. In these situations a combination of technologies or treatment stages may provide a solution (e.g. a combination of coagulation and filtration plus, where necessary, ultraviolet irradiation (UV) or ozone is often used to deal with significant *Giardia* and *Cryptosporidium* hazards in raw water). It is also important to recognise that there are some hazards which rely primarily on only one treatment stage for their removal. In such situations the reliability of that treatment stage becomes even more important.

However, irrespective of the treatment combination used, operators need to understand that there is a strong inter-dependence between the level of different hazards in the raw water, the operational performance of each treatment stage, the knock on effect to subsequent treatment stages, and potential for water quality changes to occur within the distribution network and consumers' own plumbing systems. Taking this into account, it is necessary to identify the correct process limits i.e. the maximum or minimum control limit for each stage of the process which maintains the effectiveness of that treatment stage, without compromising the next treatment process (Sancho 2009).

5.3 ABSTRACTION MANAGEMENT

5.3.1 Water quality risks within the catchment

Chapter 4 has dealt in some detail with the various hazards that can arise within the catchment and how these can be mitigated. Depending on local circumstances and the effectiveness of catchment management controls, the extent to which raw water pollution is minimised can, in practice, be very variable, particularly when water utilities do not own the catchment and/or have no direct powers to control pollution. Experience shows that the larger and more complex the catchment, the more difficult it is to ensure completely effective catchment protection. Water suppliers will therefore need to realistically assess the likelihood that pollution risks in the catchment can be controlled effectively in this way alone. Although catchment protection is by far the best option, where this is not possible within a reasonable timescale, then additional treatment or other measures to provide a suitable barrier may need to be considered in order to reliably maintain treated water quality.

Due to these factors it is important to understand the hazards in relation to the treatment capability available, enabling informed decisions to be made regarding the quality of raw water that is abstracted. Additionally, unexpected events within the catchment can create a risk of accidental pollution. Emergency procedures will thus also need to plan for unexpected deterioration in raw water quality, when existing treatment barriers may become inadequate (see Chapter 4.2.8).

5.3.2 Risk based management of abstractions

Depending on local circumstances and available water resources/treatment infrastructure, the way that water abstractions are managed can help to provide additional mitigation measures against hazards in the raw water.

For example, where multiple raw water sources are available to supply the treatment works (separate wells/boreholes or several intakes from separate surface water catchments), if one abstraction becomes contaminated, then the output from the affected abstraction may be reduced or stopped altogether, at least for a short period of time. Similarly, if raw water storage facilities are available (e.g. bankside storage reservoirs for a surface water abstraction from a river), then it might be possible to temporarily suspend abstraction from the normal source and rely on the stored water for a period. Even where only single sources exist, then reliance on treated water storage at the works and within the distribution network may provide the opportunity to temporarily suspend abstraction, or at least to reduce abstraction to a minimum during periods of exceptional raw water quality risk.

Emergency procedures developed as part of the WSP should therefore, where appropriate, include plans to manage raw water abstractions in this way. To be most effective they must be linked to good raw water quality monitoring facilities and will need to balance the risks between maintaining water supply to consumers and abstracting water of a lower quality than normal.

Raw water monitoring will need to be tailored to the risks of the catchment and the types of treatment barriers available, but Table 5.1 provides a general guide to the types of continuous (on-line) monitoring that could be useful in managing abstractions. Table 5.2 gives examples of parameters that may be useful to undertake offline at a laboratory in order to characterise the raw water quality (see also Chapter 4.2.7).

Table 5.1 Examples of parameters used for on-line raw water monitoring

Type of Source	Potential On-Line Monitoring
Groundwater	Water level relative to pump Turbidity (borehole collapse/surface water ingress) Conductivity (saline infiltration, general pollution)
Impounded surface water (e.g. impounded reservoirs/pumped bankside storage)	Reservoir level relative to draw off points Turbidity and colour pH, dissolved oxygen, ammonia (particularly if eutrophic)
Running surface water (e.g. rivers)	Flow and level of the main river flow Colour and turbidity pH and conductivity Ammonia and dissolved oxygen

Table 5.2 Examples of parameters used for laboratory based raw water monitoring

Parameter	Characterisation	Frequency of control
Raw surface water		
Metals (depending on geological features)	Only for relevant hazards	Low
Total Organic Carbon	Indirect hazard (e.g. through disinfection by-products)	Medium
Other organic chemical compounds e.g. pesticides	Only for relevant hazards	Low
Escherichia coli	Indicator of faecal contamination	High
Clostridium perfringens	Indicator of remote faecal contamination	High
Enterococci	Indicator of animal faecal contamination	High
Cyanobacteria (or other algal characterisation)	Eutrophication	Seasonal
Other organisms	Only for relevant hazards	Low/medium
Raw ground water		
Metals (depending on geological features)	Only for relevant hazards	Low
Nitrate and nitrite	Hazard	Medium
Chloride	Hazard (consumer satisfaction)	Medium
Other ions	Hazard (consumer satisfaction)	Medium
Escherichia coli	Indicator of faecal contamination	High
Clostridium perfringens	Indicator of remote faecal contamination	High

5.3.3 Raw water screening and straining

Typically, when water is abstracted from running surface water or when raw water transport is in an open channel, some form of coarse screening is required to prevent large objects blocking pipe work and restricting the flow of water to treatment. Finer screens often follow to prevent smaller objects becoming entrained in pumps in order to improve their reliability and longevity. Strainers and microstrainers may also be necessary to reduce solids which may accumulate in downstream tanks, may cause impact damage to UV sleeves in a groundwater situations or in the case of algae may overwhelm or cause significant problems with traditional clarification and filtration processes. The risks, and therefore the need for these simple mitigation measures, can be readily assessed with even a fairly basic understanding of the quality of raw water normally abstracted.

5.3.4 Risks associated with raw water storage

A well designed and maintained raw water storage facility with appropriate management procedures should not cause any deterioration in raw water quality. Indeed, in many situations storage of raw water before treatment can improve water quality, for example by reducing turbidity. Storage may also provide a valuable reserve of water when abstraction is undesirable or not possible. It can also be used as an opportunity for blending of variable quality sources therefore providing a more stable quality of water for subsequent treatment. Equally however, in some circumstances there may be a risk of water quality deterioration within stored water and this needs to be taken into account as part of the WSP.

Management of raw water storage facilities should take account of the following factors in order to assess what actions may be necessary to minimise the risk of water quality deterioration.

- The type of raw water being stored and whether settlement, oxidation or blending is to be encouraged or not.
- If it is covered or uncovered.
- The residence time and flow characteristics (plug flow, mixing, propensity for areas of low flow or stagnation).
- Any hazards or hazardous activities within the catchment including use of the reservoir for recreation.

Where practicable, small raw water storage tanks should ideally be covered, since this provides some protection against unauthorised access by people or animals and environmental effects (e.g. floods or sunlight). Exposure to sunlight can sometimes promote the proliferation of a large spectrum of organisms, especially algae, cyanobacteria, plankton and waterborne animals, particularly if residence times are long and nutrients are present in the raw water.

For raw waters with high turbidity or suspended solids, raw water storage facilities with a suitable residence time, which reduces the velocity of water flowing through it, will begin the settling or solid/liquid separation process, which will in turn reduce the challenge on downstream processes. However, regular cleaning programmes and a stilling chamber or careful control of flow changes will need to be considered to reduce the risk of settled solids being re-suspended or the development of septic or anaerobic sludge.

Raw water storage can be designed with cascades or more sophisticated aeration facilities to raise the level of dissolved oxygen and start the oxidation process to precipitate dissolved metals or displace dissolved carbon dioxide and thereby elevate pH where appropriate.

Many larger raw water storage reservoirs are too big to be covered and thus other approaches to managing the risk of quality deterioration need to be considered. Both eutrophication and thermal stratification (the formation of layers of different temperatures in a lake or reservoir) can occur in large impoundment reservoirs, which as well as creating a risk of water quality deterioration can also negate the benefits provided by settlement. The risk of short circuiting should therefore be assessed under all abstraction and raw water conditions. Eutrophication can lead to additional biomass and taste and odour problems caused by the proliferation of algae, which can affect downstream processes. After periods of prolonged stratification, significant increases in metals and nutrients can occur when "*turn over*" of the reservoir takes place in the autumn as shown in Figure 5.5 below.

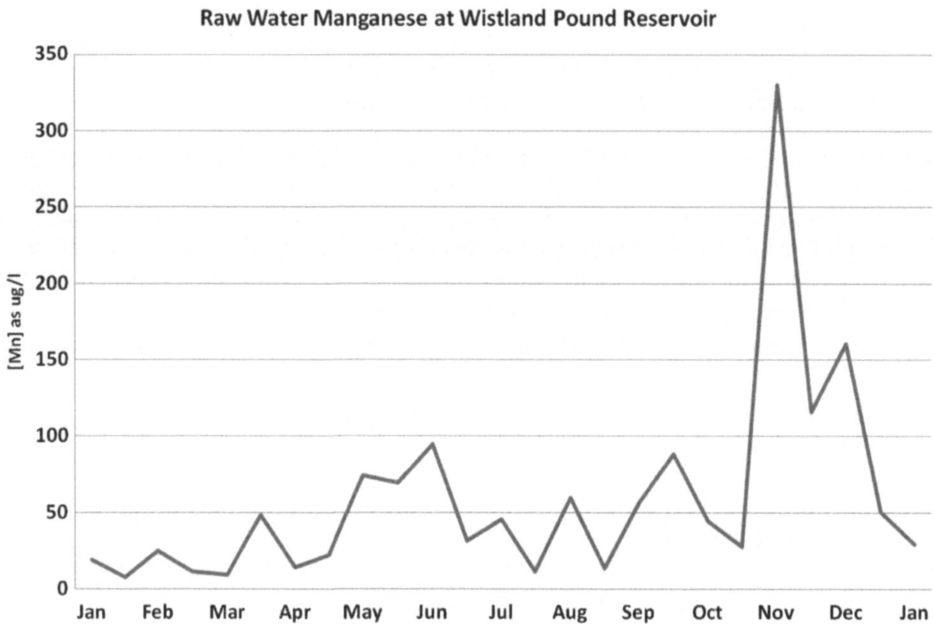

Figure 5.5 An increase in raw water manganese levels due to a late autumn turnover of a stratified reservoir

Potential mitigation measures to minimise the impacts of eutrophication include:

- Reducing the nutrient (particularly phosphorus) input upstream and into the reservoir.
- The use of submerged barley straw can help to mitigate adverse impact of algal blooms if properly installed around surface abstraction points (Everall and Lees 1996 & 1997, Garbett 2005, Murray et al. 2010).
- Regular operation of scour valves to minimise the build up of sediments.
- Where recreational fishing is practiced limiting the use of "*ground bait*" by coarse fishermen and managing fish stocks effectively.

Potential mitigation measures to minimise the impacts of thermal stratification include:

- Utilisation of varying draw off levels from within the water body to abstract the best available water for treatment.
- Artificial destratification using mechanical mixing, air lines, or helixors to prevent anoxic conditions re-dissolving nutrients and metals from the reservoir sediments.

The potential of a reservoir to stratify and subsequently "*turn over*" following stratification could be assessed by the measurement of temperature and dissolved oxygen at varying depths. This in turn can be used as a trigger to initiate destratification techniques which may need to run for as much as several hours each day to prevent thermal stratification developing. Typically, such operations cause the localised mixing of cold water from depth with the upper warmer water. Some systems of higher energy are designed to completely maintain an oxygenated sediment water interface at the bottom of the reservoir in order to minimize the recycling of nutrients and eventually break the eutrophic cycle (Hendry 2006). Other lower energy systems, which are operated intermittently, are designed to simply mix the water, and by increasing dissolved oxygen levels at depth, thus reduce the dissolution of metals and nutrients. These systems are mainly aimed at preventing or overcoming stratification.

5.3.5 Wellhead protection for groundwater abstractions

As identified in Chapter 4, management and protection of the wellhead, the borehole itself, and the immediate collection areas around boreholes and springs are vitally important activities which must be included as part of a WSP. If implemented properly these can provide relatively low cost and effective mitigation measures to prevent contamination of the raw water and are particularly important when relatively simple treatment plant is installed. Risks can occur, particularly from microbiological contaminants, as well as chemicals (e.g. pesticides or fertilisers) and fuels that might be used or stored in the vicinity of the borehole.

Preventing the direct ingress of contaminants into the top of the borehole is essential to preserve the general quality of groundwater and in particular to maintain the microbiological quality. Asset management programmes should include regular assessment of the condition and integrity of the borehole and any linings or casings that may have been installed. The design of the wellhead should ensure that rising pipework is brought well above ground level, as shown below (Figure 5.6), or by creating a bunded area around the wellhead. Where reliance on a bunded area is the preferred option then a suitable means of detecting flooding events and means of pumping away the collected water needs to be considered.

Source protection areas (see Chapter 4) should delineate the immediate area of the wellhead (zone A) and the inner catchment area (zone B) representing approximately 50 day travel time to the abstraction which is normally considered the minimum necessary for degradation of microbiological pollutants. The specific criteria which are used may vary from country to country, but the principle of giving time for biological pollutants to degrade is the same. Within these zones strict controls should be maintained on any activity which could result in contamination of the raw water. It is particularly important to manage microbiological risks from collection and disposal of human and animal wastes. Mitigation measures could include fencing to exclude livestock and relocating or improving cess pits/latrines and sewer systems. The storage and/or use of chemicals and fuels will also need to be subject to controls to minimise the risk of leaks, spills or other activity which could result in chemical contamination of the raw water.

5.4 TREATMENT CHEMICALS

Treatment chemicals are an essential component of many different water treatment processes, but must be used with care. Experience has shown that a number of potential risks exist which could lead to water contamination and/or adversely affect the efficiency of the treatment process. These must be considered as part of a WSP development. Figure 5.7 summarises risks which need to be assessed, and which are then briefly described below, along with potential mitigation options.

Figure 5.6 Typical well designed borehole headworks

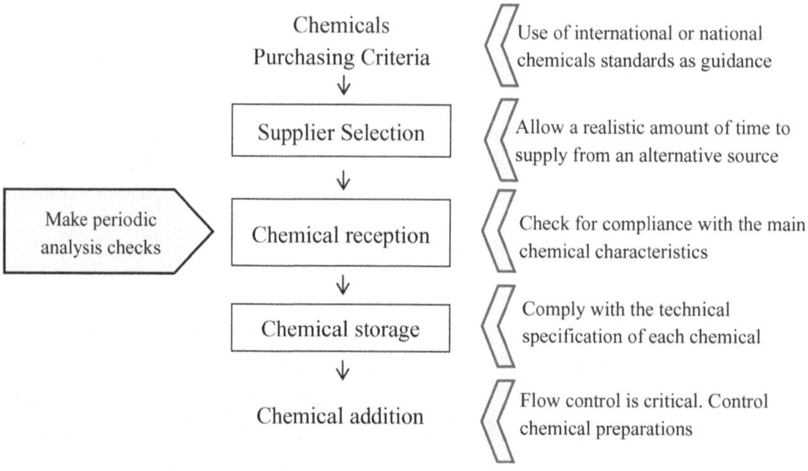

Figure 5.7 Main risks associated with treatment chemicals and mitigation measures

5.4.1 Legislation covering treatment chemicals

Many countries have local laws which control the use of certain treatment chemicals or processes. Additionally, some chemicals must meet local requirements in terms of maximum allowable contaminant and/or dose levels. Without appropriate procedures the regulatory requirements may not be met. Potential mitigation measures include:

- Ensuring that the utility has procedures to check up to date regulatory requirements covering use of treatment chemicals.
- Ensuring that operators have clear and up to date instructions about any legal requirements relating to use of treatment chemicals, e.g. maximum chemical dose to be used.
- Ensuring that chemical supply contracts include specific requirements to meet prescribed purity and maximum contaminant levels and that manufacturer puts in place measures to ensure that the chemical specification is consistently met. Ideally there should be independent verification of the purity of supplied chemicals.

Suitable control points would typically include checks to ensure operator and contractual procedures are adequate and up to date.

5.4.2 Availability of chemicals

The availability of chemicals depends critically on reliable manufacturing plants and delivery arrangements. If these fail then the supply of chemicals can be interrupted and treatment plant performance put at risk. Potential mitigation measures include:

- Assessing the reliability of the chemical supply chain and putting in place suitable controls, including for example, ensuring adequate storage to last during periods of adverse weather when supply may be more difficult.
- Implementing policies which define the appropriate amount of chemicals to store on site, taking into account the reliability of the supply chain. This needs to include such factors as rate of use, reliability of supply, and stability of chemical, as well as the cost and difficulty of storage.
- For critical chemicals, it may be necessary to have in place arrangements with more than one supplier/deliverer or ensure that alternative supply arrangements can be put in place quickly if the normal supplier has problems.

Suitable control points would typically include appropriate checks to ensure that satisfactory supply chain arrangements are in place.

5.4.3 Chemical delivery to site

Treatment chemicals may be contaminated if they are delivered in tankers or containers that have previously been used for other unsuitable products. Additionally, unsupervised deliveries to waterworks may result in a chemical being put into the wrong storage vessel at the waterworks or delivered in a way that creates health and safety or environmental risks. Potential mitigation measures include:

- Ensuring that the delivery contract includes suitable requirements in terms of cleanliness of delivery vessels and/or prior use with other products.

- Ensuring that all deliveries are supervised and that operators are given appropriate procedures to check that the delivery meets contract conditions, for example in terms of type/cleanliness of delivery vessel.
- Ensuring that the physical and procedural delivery arrangements on site are properly designed to minimise risk (e.g. lockable and clearly labelled delivery points).
- Encouraging suppliers to enter into a shared responsibility and mutually agreed procedures for the safe delivery of chemicals to protect both parties.

Suitable control points would typically include checks to ensure contractual and operator procedures are adequate and up to date. Subject to appropriate health and safety considerations, this could also include verification of the quality of certain bulk chemical deliveries prior to discharge into local storage tanks. This may be carried out, for example, by simple on-site tests such appearance, pH, specific gravity, presence of a particular chemical parameter (e.g. aluminium, iron or chlorine), or absence of objectionable odours.

5.4.4 On-site generation of treatment chemicals

Some chemicals can only be made on site e.g. ozone and chlorine dioxide or, optionally, they may be made on site e.g. sodium hypochlorite. This can reduce problems of non availability due to unreliability of external manufacturers, but in turn can create its own hazards. The quality of the chemicals used for on-site generation is also important to provide an end product of suitable quality. For example, the salt used for on-site electrolytic generation of chlorine should have a low bromide content to prevent excessive bromate formation in the hypochlorite.

The main risk mitigation options are to assess the reliability of on-site generation and ensure that appropriate back up arrangements are in place if the process fails, for example by use of standby equipment. Proper specification of the quality of chemicals used is also necessary. Suitable control points would typically include on-line monitoring of appropriate parameters to check the continued satisfactory operation of the chemical generation process.

5.4.5 Chemical storage

Some chemicals will degrade over time, leading to a decrease in effective strength or increase in the level of contaminants. This can be made worse if inappropriate storage vessels are used. Storage vessels may corrode, leading to loss of chemical, thus impacting treatment and/or creating an environmental or health and safety risk. Potential mitigation measures include:

- Ensuring that storage vessels are properly designed to minimise the rate of chemical degradation using the manufacturer and/or other expert advice. Storage vessels should also be designed to minimise the risk of vessel corrosion and appropriate arrangements should be in place to contain chemical loss if corrosion does occur (Figure 5.8).
- Procedures should be put in place to check the integrity of storage vessels at appropriate intervals.

A typical control point could include regular checks of the effective strength/ level of contaminants in stored chemicals.

Figure 5.8 An example of liquid chlorine storage

5.4.6 Chemical dosing pumps and dosing lines

Pump failure or the blockage or damage to the dosing line may result in the chemical not being reliably delivered to the appropriate part of the treatment process. Potential mitigation options include:

- Ensuring that dosing pumps are suitable for the particular chemical and, where necessary, have back up standby pump/delivery pipework if the main pumps fail.
- Ensuring that dosing lines are made of appropriate material and in locations where they can be inspected at regular intervals for damage.
- Introducing measures to prevent and/or quickly detect leaks. For example, providing loading valves or bursting discs with a discharge to an alarmed "*catch pot*" to identify blockages and avoid excessive backpressures being generated and thereby causing damage to the dosing system as a whole.

Suitable control points would typically include appropriate inspections to check that dosing pumps and lines are operating correctly, and flow alarms or measurement at or near the point of delivery into the process.

5.5 DISINFECTION

5.5.1 Introduction

Disinfection is a critically important treatment process for all surface water sources and the vast majority of groundwater sources. The primary objective is to remove or inactivate pathogenic microorganisms. More information on specific microbiological hazards is available from WHO (2011). The specific objective(s) of disinfection is usually covered by local legislation and/or advice from health authorities, but typically involves the complete absence after treatment of indicator organisms, such as *E.coli* and total coliforms, and in some cases requirements for other specific organisms, such as *cryptosporidium*, may be specified. Some countries may also specify particular minimum disinfection regimes for different types of raw water.

The most common form of disinfection is though the use of chemical disinfectants such as chlorine, chlorine dioxide and ozone. Increasingly, ultra-violet irradiation (UV) is being used as a disinfection process for many types of water. Additionally, the advent of improved and more cost effective membrane treatment processes means that, subject to suitable design and operation, this can now also be considered as providing an effective disinfection barrier for some types of water through the physical removal of pathogenic organisms.

Chemical disinfection can be applied at an initial, intermediate or final disinfection treatment stage. Primary disinfection is defined as the main process barrier against microorganisms and typically takes place in contact tanks (for chlorine, chlorine dioxide and ozone). Secondary disinfection is the process which provides a residual disinfectant to protect water during passage through the distribution network (see Chapter 6).

Treatment effectiveness for chemical disinfection is a function of disinfectant dose (after any oxidation demand has been satisfied), contact time, temperature and, for chlorine, pH. The relationship between disinfectant concentration and contact time is integral to the understanding of disinfection kinetics and the practical application of the CT concept. This is defined as the product of disinfectant concentration C (mg/l) and contact time T (minutes). The precise choice of CT for each plant will depend on local factors, including the raw water quality and the level and type of pathogens that might be present (Le Chevallier and Au 2004).

To ensure the effectiveness of disinfection, in many situations it is necessary to first remove or reduce the occurrence or concentration of other contaminants in the raw water (see Section 5.6 below). In general this requires treatment processes for the removal of suspended solids, colour and natural organic matter, the precipitation or oxidation of metals and the destruction of ammonia. Whilst these treatment processes are normally associated with surface waters, they may also be necessary for some groundwaters in specific situations. The preparation of water for subsequent disinfection is important not only in terms of the efficacy of the disinfection method employed but also in the control of undesirable by-products, many of which may have negative health or acceptability implications.

The water quality risks associated with disinfection and which need to be considered as part of a water safety plan typically fall into the following main categories:

- The microbial load in the raw water increases above the level that can be removed by the existing disinfection barriers, or the type of microbial load changes, e.g. increased levels of *cryptosporidium* which the plant was not designed to cope with (see Chapter 4).
- The effectiveness of the existing disinfection barriers reduces, e.g. due to failure or poor maintenance of treatment plant, treatment processes or units being out of commission for maintenance, or an unrecognised increase in the raw water turbidity or oxidation demand.
- The formation of disinfection byproducts. Such byproducts are often covered by legislation, but the specific requirements will vary from country to country. Although it is important that the level of byproducts is kept to a minimum, WHO advises that in seeking to minimise the level of disinfection byproducts, disinfection itself should never be compromised (WHO 2011).
- Chemical disinfectants are also oxidants to varying degrees, which therefore have the potential to produce assimilable organic carbon (AOC) which can increase the risk of biofilm formation within the distribution network if not reduced or controlled (Van der Kooij 2003).
- Non availability or contamination of the chemical disinfectant (see Chapter 5.4).

The following sections briefly describe water quality risks associated with specific disinfectants and potential mitigation options.

5.5.2 Chlorine

Chlorine is the most commonly used chemical disinfectant and can be supplied as a liquefied gas, hypochlorite salts, or can be generated on site. Although chlorine is a very good and widely used disinfectant it is not effective against all microorganisms. Secondary disinfection with chlorine also has the ability to provide a residual in the distribution system. Generally, microorganisms which have protective cell walls and/or the capacity to develop sporulated forms can be less sensitive to the application of chlorine. Independently of the supplied form, the key issues are the purity of the product and to know the concentration of active chlorine being maintained during all the treatment process in order to ensure effective disinfection. Typically, doses in the range 0.5–2.0 mg/L are used to achieve a suitable CT value, but this will vary according to local circumstances.

The main water quality risks associated with chlorine disinfection and typical mitigation measures fall into the following categories:

Reduction in disinfection effectiveness

The disinfection efficiency of chlorine depends critically on a range of factors including the chemical quality of the water being disinfected (e.g. pH, turbidity, organic content, ammonia), temperature and the CT value that is required. For each plant, disinfection conditions to ensure effective control of microorganisms need to be defined, monitored and kept under review. Any change to any critical factor, including adverse change in raw water quality, should be investigated immediately to ensure that disinfection effectiveness is being maintained.

In particular, it is important to monitor the level of chlorine at key points (which may be termed critical control points – see Chapter 3.2.5) in the process, wherever possible on a continual basis using on-line instrumentation, to ensure that the required CT is maintained. Alarms should be set for both high and low chlorine levels and procedures for rapid response to alarms should be in place. Any evidence that disinfection effectiveness (log reduction efficiency) is being compromised, for example rising levels of microorganisms or occasional detection of indicator organisms in the treated water, should always be investigated to identify possible causes.

A good understanding of the effective chlorine concentration (e.g. concentration of HOCl which diminishes rapidly at higher pH values) and the plug flow efficiency within the contact zone is important when considering the disinfection goals of a particular site. Within the primary disinfection zone, regular checks of the integrity of contact tanks and treated water storage tanks should be undertaken to ensure that no ingress of surface water or other forms of contamination (e.g. animals) is occurring. Measurement of plug flow efficiency can be made by tracer studies or by the use of computational fluid dynamics (CFD) assessments if necessary at the design stage. Where appropriate, simple modifications to inlet arrangements can dramatically increase the actual contact times, thus minimising the risk of short circuiting or insufficient disinfection due to a loss of storage volume. For example, water can be forced to travel up and down a number of lanes whose depth and width dimensions are small relative to their length.

Purchasing contracts should ensure that the chlorine disinfectant supplied by the manufacturer has the required concentration of available chlorine and also meets necessary purity levels in respect of potential trace contaminants. In some countries the actual reliability of chlorine supplies may be an issue and thus the WSP may need to address this as a key risk factor (see Section 5.4.2 above).

Disinfection byproducts

The use of chlorine disinfection can lead to the formation of a range of disinfection byproducts. The most common are the trihalomethanes (bromoform, bromodichloromethane, dibromochloromethane, chloroform) and haloacetic acids. Other chlorine byproducts can also be generated such as haloacetonitriles and chloral hydrate, although the likelihood of occurrence must be evaluated for each site. The formation of chlorine byproducts depends mainly on the organic carbon (TOC) content of the water, the chlorine dose used and the pH. Water with a TOC content >2mg/L has a higher probability of producing unacceptable levels of chlorine byproducts (Rosa 2009). Ways of minimising the level of byproducts will depend very much on the quality of raw water being treated and the configuration of the plant. Ideally, chlorine dosing should be as late in the process chain as practicable, after as much of the organic material as possible has been removed. The water treated by chlorine should have as low a level of turbidity and TOC as possible. Maximum levels of chlorine dosing should be defined which ensures adequate disinfection without generating unacceptably high levels of disinfection byproducts.

5.5.3 Chlorine dioxide

Chlorine dioxide has more oxidation potential than chlorine. Some resistant cell walled microorganisms can be inactivated by this oxidant as well as sporulated bacterial forms. However, since chlorine dioxide is very unstable, it must be generated *in situ*. Usually a stream of sodium chlorite is mixed with aqueous chlorine or hydrochloric acid under carefully controlled conditions. Although chlorine dioxide can be temporarily stored (<30min), immediate application must occur to reduce degradation of the oxidant and to avoid loss of disinfection effectiveness. Typically doses in the range 0.5–2.0 mg/L are used, but this will vary according to local circumstances.

The main water quality risks associated with chlorine dioxide disinfection and typical mitigation measures fall into the following categories:

Reduction in disinfection effectiveness

Since chlorine dioxide is generated on site, disinfection effectiveness depends critically on the reliable operation of the on site generation plant. A key control point therefore needs to be established to monitor the continual effective operation of the plant. Measures also need to be put in place to ensure that the chemicals used to generate chlorine dioxide are reliably available (see Section 5.4.4 above).

As with chlorine disinfection, for each plant, disinfection conditions to ensure effective control of microorganisms, including the CT value that is required, need to be defined, monitored and kept under review. In particular it is important to monitor the level of chlorine dioxide at key (critical) control points in the process, wherever possible on a continual basis using on-line instrumentation, to ensure that the required CT is maintained. Any change to any critical factor, including Log reduction efficiency and adverse change in raw water quality should be immediately investigated to ensure that disinfection effectiveness is being maintained. Good contact tank design, as described for chlorine, is equally important for chlorine dioxide.

Disinfection byproducts

The use of chlorine dioxide disinfection can commonly lead to the formation of chlorite and chlorate byproducts, both of which are often covered by local legislation. Chlorites are typically produced in larger quantities compared to chlorates. The production of chlorite is highly dependent on the dose applied, the chlorine dioxide generation process used and the type of water. Mitigation measures include only using chlorine dioxide when suitable to do so relative to the raw water quality, and careful management of chlorine dioxide dosing, since the presence of chlorites is proportional to disinfectant dose.

5.5.4 Ozone

Ozone is a very unstable gas which is produced on site and which must be applied immediately after production. It is a more powerful oxidant than chlorine and chlorine dioxide, and thus can be a more effective disinfectant against microorganisms which are more resistant to chlorine based disinfectants. Ozone is also used as an oxidant for some metal and organic hazards (see Section 5.7.3 below). Ozonation processes are normally applied in initial or intermediate stages of the treatment plant. Typically, doses in the range 0.5 3.5 mg/L are used, but this will vary according to local circumstances.

The technology for ozone production and application varies considerably depending on the manufacturer and this can give rise to specific operating and maintenance requirements. Since ozone is very unstable gas, ensuring effective mixing with the water and adequate subsequent contact time is a key operational issue. Additionally, it has to be remembered that disinfection (oxidation) efficiency varies inversely with the turbidity of the water. Organic and inorganic components can also be oxidised by ozone, but the efficiency with which this occurs depends on the relation between dose, contact time and quality of the influent water.

The main water quality risks associated with ozone disinfection and typical mitigation measures fall into the following categories:

Reduction in disinfection effectiveness

Since ozone is generated on site, disinfection effectiveness depends critically on the reliable operation of the on site generation plant, the ozone mixing system and the ozone contact arrangements. A key control point therefore needs to be established to monitor that the plant continues to operate effectively. Good operational maintenance practices must be implemented to ensure the reliability of plant operation, and back up plans must be in place to deal with failure of any component including loss of power. As with chlorine and chlorine dioxide disinfection, for each plant, disinfection conditions to ensure effective control of microorganisms by ozone, including the CT value that is required, need to be defined, monitored and kept under review. In particular, it is important to monitor the level of ozone at key (critical) control points in the process, normally on a continual basis using on-line instrumentation, to ensure that the required CT is maintained. Any change to any important factor, including Log reduction efficiency and adverse change in raw water quality such as turbidity, should be immediately investigated to ensure that disinfection effectiveness is being maintained. Where ozone is generated from oxygen, measures also need to be put in place to ensure that this is reliably available and of sufficient purity.

Disinfection byproducts

In the presence of bromide, bromate can be produced. This is a byproduct of increasing concern and will need to be managed carefully to secure necessary local legislative requirements. Where elevated levels of bromide are present in the raw water, bromate levels can be minimised by careful control of ozone dose and contact time. A balance is required between disinfection (oxidation) effectiveness (ozone dose and contact time) and the creation of bromate.

Of the three main chemical disinfectants, ozone is the most powerful oxidant and therefore has the greatest potential to break down organic material into lower molecular weight assimilable organic compounds (AOC). Ozone is also often applied earlier in the treatment process, where levels of dissolved organic carbon can be relatively high. The risk of subsequent regrowth of biofilms/microorganisms is likely to be mitigated by utilising a downstream process which encourages controlled biological oxidation or degradation such as activated carbon or slow sand filtration.

Maintenance of a chlorine residual can also provide a degree of mitigation but the increase in the levels of lower molecular weight organic material can also increase the level of chlorinated disinfection byproducts, if chlorine is used as a subsequent final disinfection stage.

5.5.5 UV disinfection

UV disinfection involves passing water through a specially designed reactor which is subject to moderate to high intensity UV radiation. The most effective wavelength for biocidal activity is in the range 200–300 nm and ideally around 260 nm (DWI 2010). There is a wide range of commercially available UV equipment on the market and the most appropriate will need to be identified for the specific treatment plant and raw water quality. Similar to chemical disinfection, the disinfection effectiveness is dependent on a number of variables; the dose (intensity x time) of UV applied, the UV transmissivity of the water (the ability of UV light to pass through the water), the turbidity, which can provide shielding effects as well as reducing the UV transmissivity. Typically doses in the range 25-100 mWs/cm^2 are used, but this will vary according to local circumstances and the disinfection/target organisms or goals for the site.

Effective UV disinfection relies on a low turbidity and high transmissivity water to be most efficient and thus tends to be used mainly for high quality groundwater or highly treated surface waters. Water quality data, representative of that entering the UV reactor, should be discussed with suppliers in advance of a decision on the specification of a reactor. To support this decision making process, key parameters such as turbidity and UV transmissivity should ideally be measured on-line in order to confidently predict the most challenging quality of water passing through the reactor. Seasonal, as well as diurnal variation, in water quality and the potential for fouling of the lamps should also be considered in this assessment.

UV does not produce any residual disinfectant capacity to protect the water though the distribution network, so if this is required then a separate residual disinfection process will also need to be installed. UV radiation, in conjunction with other oxidants such as peroxide, can also be used as a treatment process to remove undesirable organic compounds (see Section 5.7.3 below).

The main water quality risks associated with UV disinfection and typical mitigation measures fall into the following categories:

Reduction in disinfection effectiveness

As with other disinfectants, conditions to ensure effective and reliable control of microorganisms by UV need to be defined, monitored and kept under review. All modern UV reactors will be validated to operate within a specified range of operating conditions that will need to be understood, suitably alarmed and rapid/automatic failsafe corrective actions put in place to ensure the desired disinfection efficacy is achieved. Good operational maintenance practices, including replacement and rotation of lamps at appropriate interval, must be implemented to ensure the reliability of plant operation, and back up plans must be in place to deal with failure of any component, including loss of power. It is also known that many lamps may have a shortened life if they are cycled on and off too frequently and this should be kept to a minimum particularly in any one day.

A number of key control points therefore need to be established to verify the continual effective operation of the plant. These could include:

- UV fluence or intensity, water flow rate, and lamp status within each reactor.
- UV transmissivity or absorbance where a "*calculated dose approach*" is employed.
- Lamp operating time and lamp replacement records/schedules and turbidity.

Any change to any important factor, including adverse change in raw water quality such as turbidity, should be investigated immediately to ensure that disinfection effectiveness is being maintained. Some raw waters, for example those containing trace amounts of iron or manganese, can result in scaling or deposits on the reactor surface and fouling lamps. This needs to be taken into account during plant design and, where necessary, arrangements made for routine cleaning without compromising ongoing disinfection capability.

Disinfection byproducts

Under certain conditions and depending on the quality of the water being treated, low levels of byproducts can be formed with UV disinfection. For example, at wavelengths less than 240 nm, nitrate in the water may be converted to nitrite (DWI 2010). The problems are usually less significant than with chemical based disinfection but do need to be assessed.

5.6 SOLIDS/LIQUIDS SEPARATION AND REMOVAL PROCESSES

5.6.1 Introduction

Particularly for surface waters, but sometimes for groundwater, production of safe drinking water requires treatment to separate and remove solid, colloidal and dissolved material from the raw water. This comprises a complex and wide range of material including:

- Organic matter arising from human activity as well as natural organic matter, both of which can promote biofilm growth within the network and also act as precursors for disinfection by products such as trihalomethanes.
- Microorganisms, including protozoal parasites such as *cryptosporidia.*
- Dissolved metals.
- Algae.
- Inorganic sediment.

Treatment to remove these substances is usually the first part of the process chain and can provide a robust barrier against high levels of colour, turbidity and microbiological quality risks. Significant improvements in downstream disinfection performance, reduced byproduct formation and improvements in water acceptability to consumers are also normally secured.

All stages that contribute to the overall solids/liquid separation process must be considered and performance must be evaluated as a sum of all the parts. The most important issue is to ensure that the overall process limits and capacities are understood for each stage. Knowledge of these stage capacity limits can allow estimates to be made about the impact of any single stage failure to the overall clarification process, and also assess the potential of subsequent processes to reduce the hazard to an acceptable level. Turbidity can be used as a good indicator to evaluate clarification capacity.

Surface water treatment usually entails coagulation/flocculation followed by clarification and filtration. Direct filtration of coagulated waters can only be employed for relatively good quality surface waters where the risks are relatively low or the filtration capacity is large. Some groundwater also requires pre-treatment, for example, to remove solids or dissolved material such as metals. This section briefly describes these processes and factors that need to be taken in to account when developing a treatment WSP.

5.6.2 Coagulation/flocculation

Coagulation and flocculation are processes that are typically used to enhance the effectiveness of subsequent solid/liquid separation processes. Normally iron or aluminium salts are employed to bring together and entrain small particles, destabilise colloidal material by reacting with natural organic matter, improve organic colour removal and co-precipitate a variety of metals by the formation of metal hydroxide or polymeric flocs. Usually iron or aluminium coagulants are added to the water in a highly stirred tank and pH corrected to a point where precipitation (floc formation) occurs. In some instances, such as iron rich groundwaters, the adjustment of pH alone can promote the formation of flocs. In other situations the use of polymeric coagulants may be useful as they can negate the need for rigorous pH correction or enhance subsequent separation performance.

In order to manage coagulation processes effectively it is important to determine the optimum coagulant dose and pH for the range of raw water conditions typically experienced. In general, the main coagulant demand will come from colour, dissolved organic matter or turbidity, and the optimum coagulation conditions will depend on the treated water quality goals that have been set. For stable waters, such as impounding reservoirs, occasional manual adjustments to coagulant dose and pH may be acceptable as conditions gradually change throughout the year. However, it is important to recognise that more variable sources, such as rivers, may require several adjustments in a short period of time, which may only practically be achieved by some form of automated control system.

Water quality risks and their mitigation

The main risk to water quality derives from poor design and/or operational control of the coagulation/flocculation process. Controlling coagulation chemistry in terms of coagulant dose and coagulation pH are the two most critical factors. Under-dosing of coagulant or failing to maintain the coagulation pH within the metal solubility range will result in elevated dissolved iron or aluminium in the treated water. Under-dosing of coagulant will also result in the incomplete removal of colour and dissolved organic matter or turbidity which will impact on the effectiveness of subsequent processes. Over-dosing of coagulants can lead to high concentrations of residual salt being carried over into the clarified water, thus challenging the downstream filtration stage and ultimately increasing the risk to treated water quality.

Where polymeric flocculent aids (e.g. polyacrylamide) are used to enhance floc settlement rate, care must be taken to ensure that they are used at dose rates which do not create undesirable levels of impurities (e.g. acrylamide monomer) which are often covered by local regulatory requirements. These contaminants cannot be measured easily on site, so risks are normally controlled by product specification which allows maximum dose limits to be calculated and included in on-site operational control procedures. In calculating maximum dose rates for polyelectrolyte, any use in sludge treatment where supernatant waters are returned to the plant inlet also need to be taken into account.

Often, correct addition and control of coagulant chemicals can be compromised by poor physical engineering of the treatment plant. This can occur either through ineffective mixing at the point of chemical addition or by the destruction of flocs by excessive shear forces on transfer from the flocculator to clarifier or filters. Good design of structures and mixers should mitigate these risks. Additionally, it is important to allow sufficient time for flocs to form. For example, the time required is often significantly increased at low temperatures, low carbonate, low colour and turbidity which can occur in upland sources in winter months. As carbonate is a critical factor in the coagulation reaction, the addition of alkali before the addition of acidic coagulants can help fix free carbon dioxide (as carbonate) and improve coagulation performance under such cold clear conditions.

Effective mitigation measures for the coagulation/flocculation process include:

- Appropriate selection of coagulation chemicals.
- On-line measurement of the primary factors impacting coagulation, e.g. colour/DOC, turbidity, in order to determine and maintain the optimum coagulant and flocculent doses.

- Control of the coagulation pH and coagulant dose of the dosed water, perhaps linked to alarms to indicate potential failures.
- Effective maintenance of the dosing systems and the provision of duty/standby pump arrangements.
- Regular visual inspection of the operational performance and/or laboratory assays for coagulant and flocculent doses, in order to avoid insufficient or excess chemical doses.
- Effective design of the flocculator to ensure sufficient mixing at the point of chemical addition and avoid excessive shear during flocculation.

Suitable key (critical) control points for this process could include coagulant dose, coagulation pH or soluble iron/aluminium, where necessary measured by on-line instrumentation. Downstream measurements of clarified turbidity, clarified colour or dissolved organic material can also be used as secondary control measures to demonstrate that the coagulation and flocculation processes are optimised.

5.6.3 Clarification

Following coagulation/flocculation, water usually passes through some form of liquid/solid separation stage using a sedimentation or flotation clarification process to remove the majority of the floc that has been formed. The performance of the clarification process depends critically on the quality of flocculated water and is a pivotal stage in the performance of downstream processes.

There are a wide range of sedimentation/clarification technologies and each has to be carefully designed according to the type of water being treated and other practical considerations. In some cases this might be a simple sedimentation basin, but more typically this involves a sludge blanket clarifier or dissolved air flotation. The choice of sedimentator/clarifier is a matter for local engineering judgement, but in all cases the main issues are to keep flow conditions as stable as possible, ensure effective sludge removal by avoiding short circuits and dead zones, and to ensure a proper balance between water flow and sludge removal, in order to maintain optimum hydraulic conditions for sedimentation.

Water quality risks and their mitigation

The main risk to water quality derives from poor design and/or operational control of the sedimentation/clarification process. These include:

- Inadequate flow distribution between process units, or rapid flow changes, leading to potential hydraulic overloading.
- Peak plant output coinciding with periods of worst raw water quality, thus compromising clarifier performance.
- Streaming or bypass of the sludge blanket, which increases floc penetration to the subsequent filtration stage.
- Long sludge residence times, which can lead to microbiological activity and disruption of sludge blankets, as well as the formation of septic sludges.
- Ineffective sludge removal.
- In tanks which are open to the atmosphere some problems can occur due to sunlight increasing algal growths within the sludge and/or wind and rain can cause disruption to clarification performance.

The main mitigation options involve careful management of flow to ensure optimum hydraulic performance coupled with regular procedures for removal of accumulated sludge. Where tanks are open, consideration should be given to covering to minimise the risk of algal growths or weather disrupting clarifier performance.

Regular monitoring of the preceding coagulation/flocculation stages should also take place to ensure that optimum floc quality is passing through to the clarification stage. In general many clarification processes perform best under relatively stable flow conditions, so demand forecasting and the use of downstream storage to predict and balance changes in demand will reduce the risks associated with hydraulic shock.

Typical control points normally involve monitoring the turbidity and/or colour/dissolved organic carbon level of the clarified water, ideally using on-line monitors. Additionally, flow monitoring could also be used as a control point to measure actual against design hydraulic performance, and checks made to ensure regular removal of sludge against agreed operational performance criteria.

5.6.4 Filtration

Filtration is an important part of the solids separation treatment process, and for surface water is often one of the last stages to reduce or eliminate hazards prior to final disinfection (Bouchier et al. 1998). There are a variety of filtration technologies available. The most common is probably rapid gravity filtration where water is fed into the top of the filter and allowed to permeate through the filter media by gravity. In some situations filtration may be carried out under pressure in sealed vessels. The filtered water is then removed via under-drains to the next part of the treatment process. The choice of filter media is important. Typically, the main media is sand supported on a bed of gravel, but other suitable material can also be used. In some situations dual or even triple media filters are used. Periodically the filter media must be backwashed to remove the trapped material, thus maintaining filtration efficiency. In some situations slow sand filters can be used which have much slower filtration velocities and where filtration takes place mainly in a biologically active upper layer known as the "*schmutzdecke*".

Water quality risks and their mitigation

Although filters are a key barrier against many important water contaminants, depending on the precise design, there are a number of risks associated with their operation which could compromise their effectiveness and which needs to be managed. These include:

- Hydraulic overloading - it is important to check that filters continue to operate within their design maximum loading rates or within loading rates that are subsequently shown to be optimal for filter performance. Allowance must be made for increased loading on remaining filters when one or more is out of service for backwashing. Particular risks can occur under abnormal situations, e.g. when plant output has to be increased to cope with unexpected demands. Where multiple filters operate in parallel, the hydraulic distribution between individual filters should be balanced so that overloading of one particular filter does not occur. Where possible, the rate of flow change through a filter should be controlled to low percentage of its design flow.
- Filter media loss/damage - over a period of time filter media can be lost, reduced in size through attrition, or its effectiveness compromised by, for example, poor backwashing. Both the depth and integrity of filter media should be checked periodically, along with an assessment of the wash pattern and any evidence of dead areas or "*mud balling*" investigated.
- Ineffective backwash arrangements - regular backwashing to remove accumulated particulate material is vitally important (see Figure 5.9). The frequency depends on local circumstances, but is often triggered by headloss across the filter or increasing turbidity on filter outlet. However, washing at planned time intervals prior to headloss or turbidity triggers being reached is preferred. Many filters also use air as well as water for backwashing, since this improves removal of accumulated debris and prevents formation of so called "*mudballs*" within the filter media. Periodic checks should be made of the efficiency of air blowers.

- When filters are returned to service after backwashing, the first few bed volumes of filtrate are often of poorer quality. In some countries it is therefore common practice to adopt a "*slow start*" approach to returning filters to service and in some cases the first few bed volumes of filtrate are run to waste.
- The choice of wash water can be important, particularly where final water pH correction is practiced. Washing and re-filling with water outside the pH solubility range of the coagulant used for clarification can cause quality deterioration for the first few bed washes upon return to service.
- If significant seasonal variation in water temperature, and hence water density, is experienced, it may be necessary to adjust backwash rates accordingly to prevent carry over of sand/loss of sand during the wash cycle.
- Feed water quality deterioration - filters are normal designed to cope with certain levels of feed water quality. Typically, this might be no more than 3 NTU turbidity, but in some cases higher levels of turbidity can be adequately dealt with. If feed water quality deteriorates then the filter performance can be affected and, as a minimum, increased frequency of backwashing will be required.
- Failure of underdrains - if the underdrainage system fails for any reason, then filtration effectiveness can often be rapidly impacted.
- Failure of control valves - depending on filter design, it may be possible for failed or leaking valves to allow dirty or unfiltered water to enter the filter outlet system, particularly during the filter backwash cycle. To mitigate this risk, interlocks should be considered to prevent the incorrect sequence of valve operation, or automated wash sequences continuing, if valves fail to close or open.
- Ingress of extraneous material - filters are sometimes incorporated within buildings, but can also be outside. In this latter situation the filter surface can sometimes be contaminated by a range of extraneous material including leaves and other debris, bird droppings, insects and similar factors which might adversely impact water quality. In such circumstances it may be necessary to install some form of permanent or temporary cover.

Figure 5.9 Backwashing in rapid sand filters

The most appropriate control points to monitor the effectiveness of filter performance can only be determined locally, taking into account a range of factors. However, typical controls could include:

- Turbidity - this is one of the most common and important indicators of filter performance. WHO advises (WHO 2011) that a well operated filter can typically achieve turbidity levels of 0.1 NTU

or, ideally, less for most of the time. The use of statistical process control techniques (see Section 5.9.3 below) can be a very useful way of detecting early evidence of deterioration in filter performance. Even for plants which normally achieve very low turbidity values, any unexpected deterioration in filter turbidity should be investigated.
- Physical inspection - as indicated above, visual inspection of the filter media and backwashing operation by experienced operators can often reveal early signs of problems. Simple sieve analysis can also be used to monitor changes in media size.
- Headloss - as described above, filter headloss is often used as a trigger for filter backwashing. However, it can also be used as a potential control point, since unexpected change can indicate deterioration in upstream process or filter performance.
- Laboratory analysis targeted at the metal salt used as coagulant and removal of protozoa cysts/oocysts or spore forming bacteria.

If a plant has a number of filters operating in parallel, it is particularly important to check that any of the above risks are not just occurring in one filter. For example, if filter performance (e.g. turbidity) is only measured on the combined outlet from the whole bank of filters, then a deficiency in one filter might be missed. For this reason measuring/assessing individual filter performance is recommended.

5.7 ADVANCED/ADDITIONAL TREATMENT PROCESSES

5.7.1 Introduction

In addition to the common processes described above, many modern treatment plants also have a range of additional and/or advanced treatment processes to deal with specific hazards that might be present in the raw water. The range and complexity of technologies available is increasing all the time, so this section only briefly describes some of the more common processes and the water quality risks that may be associated with them and which need to be considered as part of a treatment WSP.

5.7.2 Carbon adsorption

Activated carbon is a treatment process which aims to remove/reduce the concentration of certain organic and, on occasions, inorganic contaminants present in the water through adsorption mechanisms. For example, it is often used to provide a barrier against taste/odour producing chemicals (e.g. those derived from algal activity in the raw water), but it may also be used to remove man made chemicals such as pesticides.

Activated carbon can be made from different carbonaceous materials, including coal, wood and coconut shells, and is manufactured with a variety of physical characteristics. Activated carbon can be used either as a powder (PAC) or in a granular form (GAC). It can remove a broad spectrum of organic material, but the adsorption characteristics of each compound will vary considerably and can be described through adsorption isotherms. Different carbons have differing capacity for different types of organic material.

PAC is normally only used to control known seasonal problems in the raw water, or as an emergency treatment to deal, for example, with accidental pollution. When PAC is used, it is normally added as a slurry before the clarification/filtration stage. However, when doing so the PAC must be fully removed by the subsequent processes. Thus the total solids loading must be considered before PAC is used to ensure that the removal capacity is adequate.

GAC is normally used in dedicated contactors downstream of the final physical barrier (filters or membranes). A range of factors influence the adsorption of different organic material, but typically those compounds which are more hydrophobic and/or larger molecular weight bind more strongly. If GAC is being used to control specific organic hazards e.g. pesticides, then it has to be remembered that these will be competing for absorption sites with other non specific organic material (e.g. natural organic matter). Eventually the adsorption capacity of the carbon will be exhausted and the risk of breakthrough, or in some cases desorption, of adsorbed material may occur. To mitigate these risks the carbon will have to be temporarily removed and regenerated in specifically designed furnaces. Calculating the necessary average bed life of the carbon is a careful balance between, on one hand, the cost and operational difficulty of regeneration, and on the other, the risk of breakthrough of the target organic compounds.

Water quality risks and their mitigation

Generally there are limited risks to water quality from well managed GAC contactors. However, a number of factors may need to be considered as part of a treatment WSP. These include:

- Careful choice of the size of GAC contactors (the empty bed contact time - EBCT) relative to the scale of barrier against organic material that is required.
- Careful choice of GAC to ensure that removal of the target organic molecules is optimised.
- Identifying the correct regeneration period for the carbon which minimises the risk of breakthrough of undesirable organic compounds. This can be done using pilot GAC filters, specially designed rapid test columns, or with advice from manufacturers, and further optimised by practical experience. The use of surrogates such as iodine or methylene blue number can also provide a guide to remaining adsorption capacity, but care should be taken as the adsorption characteristics often differ from that of the target compounds.
- Most GAC contactors also require some form of backwashing, although this does not normally require the same frequency as sand filters. Backwashing needs to be carried out carefully to avoid degradation of the GAC.
- Failure of underdrains - if the underdrainage system fails for any reason, then GAC particles can enter the final water.
- Invertebrate infestation can occur within GAC contactors and monitoring should be undertaken during the high risk periods of spring and summer. If the problem develops, it can be controlled by anaerobic conditioning, or in extreme cases, by downstream microstraining.
- If operated at low flow, after even relatively short periods off line, low dissolved oxygen/anaerobic conditions can prevail and ammonia quickly generated. Careful washing and return to service procedures should be devised following period of non use, or commissioning new or regenerated carbon.
- In some situations, particularly, for example, when preceded by ozonation, GAC can develop a biological activity within the carbon matrix. If properly controlled, this can be of value in reducing the level of biodegradable organic matter and thus the level of potential microbiological regrowth within the distribution network. However, if not controlled, this biologically active carbon can lead to unacceptable tastes and odours.

A typical control point for GAC is monitoring to ensure that the average GAC age (time between regeneration), identified as being sufficient to remove or reduce the required organic molecules, is maintained. However, this period needs to be kept under review. It is also good practice to stagger the regeneration of multiple beds over time, so that there is always a mix of GAC ages to minimise risk of breakthrough.

5.7.3 Advanced oxidation processes

Advanced oxidation processes to remove organic (or sometimes inorganic) hazards are becoming increasingly necessary as the range and level of raw water pollution risks increase, and water resource shortage, requires the use of more polluted raw water. Typically these can include:

- Ozonation for oxidation as well as disinfection purposes (see Section 5.5.4 above).
- Ozonation in combination with other oxidants such as hydrogen peroxide.
- UV (at much higher doses than used for disinfection) coupled with other oxidants such as hydrogen peroxide.

Water quality risks and their mitigation

The use of these technologies typically targets the destruction of specific organic micropollutants such as pesticides, taste and odour causing compounds (geosmin/methyl isoborneol), endocrine disruptors and pharmaceuticals. The operation of advanced treatment processes requires specialist process design and careful monitoring or piloting, so that the correct operating conditions are maintained and the identified hazard is reduced effectively. In general, the use of powerful oxidants can produce a significant increase in assimilable organic carbon (AOC) which may lead to regrowth in the downstream distribution network.

In the case of ozone, the formation of bromate is a known risk, as described in the disinfection section above. When operated in advanced oxidation mode, either alone or in combination with peroxide, bromate formation may be controlled due to the lower contact times or the presence of hydroxyl radicals. However, the higher doses potentially required may counter these mitigating effects and therefore an assessment of the risk needs to be performed.

In the case of UV in combination with peroxide, the formation and removal of genotoxic activity has been studied in full scale installations (Heringa et al 2011), and in general these risks have been mitigated by the use of downstream granular activated carbon.

The use of downstream granular activated carbon is generally advisable following all forms of advanced oxidation to reduce the by products that are commonly formed.

5.7.4 Membranes

Membrane processes are increasingly being used to meet a range of water treatment needs and this is likely to increase as the technology further improves. For example, one particular use which is becoming more significant is for desalination. Membranes are sophisticated treatment processes and need to be designed and operated in conjunction with advice from manufacturers. There are a wide variety of membrane types available which can be used to remove a range of different hazards from water.

Membranes fall broadly into two groups:

- Those which remove larger particles such as algae and protozoa, primarily microfiltration (for particles in the range 0.05–5 μm), and ultrafiltration membranes which can remove the majority of other microorganisms (in the range 0.001–0.02 μm in size) as well as some particulate matter.
- Membranes such as reverse osmosis or nanofiltration which, depending on their design and operation, can remove a wide range of much lower molecular weight substances.

Water quality risks and their mitigation

All membranes require pressure to be applied to the influent water, but this varies considerably according to the type being used. Membranes are prone to fouling and can typically require significant pre-treatment to minimise this risk. Membranes can also rupture, which reduces their effectiveness in removing identified hazards. Mitigation measures to address these risks include:

- Ensuring that influent water is of a suitable quality to minimise risk of fouling, and instituting membrane cleaning regimes as recommended by manufacturers.
- Managing pressure and flow to minimise the risk of membrane rupture.
- Having in place suitable procedures to respond to critical failure of the process, for example loss of power or pressure.

Typical control points for membranes might include:

- Monitoring influent water quality to ensure that is it suitable to minimise membrane fouling and ensuring that required cleaning regimes are maintained.
- Ensuring membrane integrity through controls such as pressure drop tests across the membrane when subjected to air injection, or through on-line particle counters on the treated water effluent.
- Monitoring trans-membrane pressure to ensure that it is maintained within manufacturers recommendations.

There are particular water quality risks when using membrane technology for desalination, including, for example, the fact that desalinated water is very low in minerals and is poorly buffered. Thus it is usually aggressive to cementitious and metal materials used in storage, distribution and plumbing systems. More information can be found in a World Health Organisation publication (WHO 2011b).

5.7.5 Ion Exchange

Ion exchange is a process used to remove a variety of dissolved ion and low molecular weight organic acids hazards from water. The treatment relies on passage of the water through a specially manufactured resin which exchanges ions present in the water for other ions bound onto the resin matrix. There are two main types; cation exchangers which are typically used for softening by removal of calcium and magnesium from the water, and anion exchangers which can remove unwanted anions such as nitrate. In some cases ion exchange resins are fairly broad spectrum whereas in other cases the resins can be more selective, for example for certain anions such as nitrate. When the capacity of the resin is exhausted it has to be regenerated by use of a solution of, for example, sodium chloride.

Water quality risks and their mitigation

When designed and operated with care, ion exchange resins can be very effective in reducing the presence of unwanted ions in the treated water. However, there are a number of risks to water quality arising from their operation which need to be taken into account as part of a treatment WSP. These include:

- Breakthrough of the ion to be removed from the water.
- Undesirable changes in the ion content of the water, for example increase in the sodium content and changes to the chloride/sulphate (Water Research Foundation 2010) or

chloride/bicarbonate ratio which can under some circumstances impact on corrosivity and lead levels at consumer taps.

Mitigation measures typically rely on approaches such as:

- Careful design of the plant and choice of ion exchange resin to ensure that undesirable changes to ionic content are minimised.
- Where appropriate, treating only part of the flow and blending the effluent with untreated water.
- Having a suitable number of ion exchange vessels operating in parallel, so that one can be out of service for regeneration without impacting on overall performance.
- Carefully designed operating procedures to manage resin regeneration before breakthrough occurs.

Control points would normally involve monitoring of key ions to ensure that target concentrations are not exceeded.

5.8 WATER CONDITIONING

5.8.1 Introduction

Before water is pumped into the distribution network it is common to further treat or "*condition*" the water to protect water quality as it passes through the network and consumer premises. This is usually either to maintain microbiological quality by use of a disinfectant residual or minimise the risk of corrosion. The WSP needs to take this into account and include the setting of targets for water quality leaving the works to ensure that water quality is maintained during distribution (see Section 5.10 below).

5.8.2 Disinfectant residual

As described in Chapter 6.7, it is normal in many parts of the world to establish a disinfectant residual for water leaving the waterworks to ensure that microbiological quality is maintained throughout the distribution network. The exception may be those situations where high quality and biologically stable water passes through short networks which are in good condition. The type and level of residual disinfectant chosen will depend on a number of local factors, particularly the length, condition and configuration of the network, and may also be covered by local regulations. It will also need to take into account the potential for disinfectant byproduct formation and the impact on consumer acceptability in relation to water taste. In some cases it may be necessary to set a lower level of residual leaving the works and boost the level further at strategic points within the network.

There are normally three main residual disinfectants that can be used (see Section 5.5 above) :

- *Chlorine*: An effective network disinfectant commonly used in many countries, but can lead to a range of disinfection byproducts (DBPs). The rate of DBP formation is influenced by pH, time, temperature and organic precursor levels.
- *Monochloramine*: Monochloramine is usually pre-formed at the water works. It persists longer in the network than chlorine and tends to form lower DBP levels although it is a less powerful disinfectant. Monochloramine formation at the waterworks must be carefully controlled to avoid formation of dichloramine and trichloramine in the network, both of which have very strong tastes which consumers find unacceptable. Care must also be taken not to mix chlorinated and chloraminated

water within the network since this can also lead to formation of adverse tastes. Nitrite can also occur in distribution as a result of chloramination.

- *Chlorine dioxide*: This is a powerful disinfectant which minimises formation of organic DBPs such as THMs, but does break down to chlorate and chlorite, both of which have been the subject of health advice from WHO and which may be subject to local regulatory requirements.

The WSP will need to take account of the type and level of disinfection residual that needs to be established and ensure that measures are in place to maintain the residual at the target concentration, and control points established to monitor the target residual against actual values. Operational procedures should be in place to react effectively if the residual level drops below the target.

5.8.3 Corrosion control

Water passing through the distribution network and consumer plumbing systems can come into contact with a wide range of different materials. The quality of final water quality leaving the works can have a significant impact on the risk of subsequent corrosion of those materials, leading to deterioration of water quality. Corrosion control is therefore an important aspect of a WSP.

Corrosion control is a complex topic and can involve many parameters, including the concentrations of calcium, bicarbonate, carbonate and dissolved oxygen, as well as pH. The detailed requirements for corrosion control differ depending on water quality and the materials used in the distribution system and consumer's premises (WHO 2011). The pH controls the solubility and rate of reaction of most of the metal species involved in corrosion reactions. It is particularly important in relation to the formation of a protective film at the metal surface. For some metals, alkalinity (carbonate and bicarbonate) and calcium (hardness) also affect corrosion rates. Corrosion strategies relevant to local circumstances therefore need to be developed to cover both materials within the distribution network and consumer plumbing systems, including:

- Unlined iron distribution pipes which can lead to water discolouration.
- In some situations care also needs to be taken where low alkalinity water passes through concrete or concrete lined pipes, since this may lead to salts leaching from the concrete thus increasing pH.
- Lead corrosion (plumbosolvency) is of particular concern where lead piping is used in consumer plumbing systems. The solubility of lead is governed by the formation of lead carbonates as pipe deposits. The solubility of lead increases markedly as the pH is reduced below 8 because of the substantial decrease in the equilibrium carbonate concentration. Thus, plumbosolvency tends to be at a maximum in waters with a low pH and low alkalinity. Control measures can include increasing the pH to 8.0–8.5 where this is practicable, or to dose orthophosphate (Hayes 2010).
- Corrosion of copper pipework and hot water cylinders can cause blue water, staining of bathroom fittings and, occasionally, taste problems. General copper corrosion is most often associated with waters with pH below 6.5 and hardness of less than 60 mg of calcium carbonate per litre. The solubility of copper is mainly a function of pH and total inorganic carbon. Solubility decreases with increase in pH, but increases with increase in concentrations of carbonate species. Control measures could include raising the pH to between 8 and 8.5 where practicable (Foundation for Water Research 2010).

5.9 WATER TREATMENT MONITORING AND CONTROL

5.9.1 Types of monitoring

A critical part of any treatment WSP is operational monitoring to ensure that treatment processes are operating as planned and within expected limits (Figure 5.10).

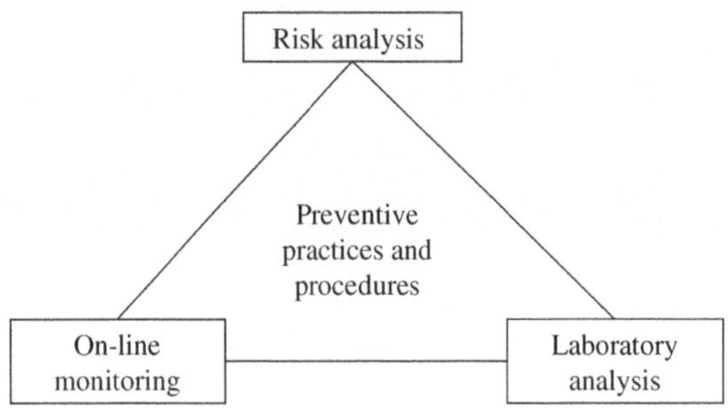

Figure 5.10 The important role of operational monitoring

Typical control points (see Chapter 3.2) for each type of treatment process have been identified in the preceding sections of this Chapter. It is vital to ensure that the control point at which each treatment stage is monitored is representative of that treatment process. Monitoring the effectiveness of treatment processes can be secured through a variety of means including:

On-line monitoring

Wherever practical, continuous on-line monitors should be used to measure performance against critical control limits. Water quality control technology is continually developing and there is a huge and diverse range of on-line instruments now available on the market (e.g. pH, turbidity, chlorine, aluminium). The selection of specific equipment and methodologies must take into account the costs involved, the expertise needed for their operation and maintenance, and the relevance of the data obtained. Sample point selection and design is also a key factor, and utilities must ensure that the water being monitored and controlled is representative of the water being treated. It must also cover the performance of the equipment in relation to factors such as resolution, limits of detection, errors, calibration and maintenance procedures.

Since on-line equipment will normally be used to monitor critical control points, it is essential that it performs reliably and within defined limits of accuracy. The equipment must be regularly maintained, according to manufacturer requirements, and regularly validated to ensure that measurement errors are within acceptable limits and that data obtained is credible.

On-site laboratory test kit equipment

Where on-line equipment is not available, portable laboratory test kits can be used for many simple process control parameters such as chlorine, turbidity or pH. Use of such equipment by operators has also been shown to improve their ownership and understanding of the water treatment process. However, where this type of equipment is used to control critical limits, a process control plan must be developed with defined monitoring frequencies and responsibilities. As with on-line monitors, the performance of portable test kits must also be properly controlled and equipment must be regularly calibrated.

Laboratory analysis

Although laboratory analysis should not normally be used to control critical process limits due to the time taken to obtain the results, in some circumstances it may be necessary where no alternative exists. Laboratory analysis may also be used for a variety of other treatment WSP purposes including:

- Assessing the concentration of critical hazards in raw water and water sources. If direct measurement is not possible, indirect parameters may be used.
- Evaluating the effectiveness of each treatment stage in reducing or eliminating a specific hazard or indicator parameter.
- Demonstrating the absence of critical hazard in treated water.
- Verifying the performance of on-line equipment.
- Confirming that the control measures taken have reduced any identified hazard to acceptable limits.

The frequency of monitoring depends on a number of variables, including the probability of occurrence, whether the data is used immediately or for medium term decision making, the type of water (raw, filtered, treated), the historical data and any statistical requirements, and costs.

Asset integrity

Ensuring that the water treatment assets are in good condition and perform effectively is an important component of operational monitoring for a WSP. Asset maintenance procedures and a long term replacement strategy should be part of an overall asset management plan (see Chapter 3.3). Regular inspection of critical treatment assets should take place, taking account of factors such as risk criticality, age, information from previous inspections and historical performance.

5.9.2 Data acquisition and quality assurance

Regular monitoring of equipment used in the treatment process is a critical part of the WSP. All the equipment that contributes to the effectiveness of a control measure, particularly those that are deemed critical, should be regularly monitored and alarmed to check that it is operating normally. Ideally this should be through remote on-line SCADA (supervisory control and data acquisition) systems (Figure 5.11) or, if not possible, manually at very regular intervals e.g. at least hourly. If on-line or regular manual monitoring is not possible, reliable alternative equipment failure warning mechanisms should be implemented, e.g. automatic SMS text messages to a control room.

Data capture and evaluation is particularly important to ensure that treatment processes continue to work optimally and are providing effective barriers to the hazards identified as part of the WSP. However, only data that can provide useful information should be captured. Data capture procedures must specify the frequency of data collection, the accuracy required, the type of statistical analysis performed, and the human or software resources used for such analysis.

Data that is captured can be used for a range of decision making purposes and also to review the appropriateness of hazard control points. Statistical analysis of data can be used to assess issues such as:

- Raw water quality trends for each parameter and hazard evaluated.
- The performance of each treatment stage against operational limits for each hazard.
- Identifying normal operating regimes for optimum treatment performance.

Figure 5.11 Example of a SCADA room

It is particularly important to ensure that all relevant data, both water quality and operational conditions, are captured when a critical operational limit deviation occurs, in order to evaluate what went wrong and what can be improved in the future.

5.9.3 Statistical process control

Statistical process control (SPC) is the application of statistical methods to the monitoring and control of a process to ensure that it operates at its full potential to produce products which conform closely to that specified. While SPC has been applied most frequently to controlling manufacturing lines, it applies equally well to any process with a measurable output, including water treatment plant. Key tools in SPC are control charts and a focus on continuous improvement.

Much of the power of SPC lies in the ability to objectively examine a process and the sources of variation in that process. In the case of water treatment plant, any variation in the process that may affect the quality of the end product can be detected and corrected early, thus minimising the likelihood that water quality problems will arise. With its emphasis on early detection and prevention of problems, SPC has an advantage over other quality methods, such as inspection, that seek to detect and correct problems after they have occurred.

5.9.4 Setting operational quality targets for treatment processes

When defining the critical limit for operational targets in treatment processes it should:

- Be clear to all the personnel involved what the limit means and what to do if limits are exceeded.
- Be realistically achievable.
- Be easily and quickly measured.
- Makes sense to the operator.

This final point is very important, because personnel must know and believe that the limits applied will really contribute to the production of safe drinking water. One way to do this is to be able to demonstrate the links between public health protection and operational and critical limits. Some of these links are easy to make because of the significant amount of available literature, for example the relationship between chlorine disinfection and inactivation of faecal indicator organism like *E.coli*. However, for other health related parameters, for example disinfection byproducts, it may be more difficult.

In some situations the involvement of local health or regulatory authorities may be useful in evaluating the health significance of each identified hazard, and the necessary operational limits to reduce the hazard to an acceptable level. However, it must be remembered that the necessary operational control limit may vary depending on factors such as the quality of raw water, the performance of each water treatment process, and the residence time.

5.9.5 Validation of operational and critical limits

It is important to validate the operational control limits to ensure that they result in the necessary health, legal and other standards being met in the treated water (Fewtrell and Bartram 2001). The challenge is to show that the limit used for the control measures is positively correlated with the absence of the identified hazard. However, this is not an easy task, and is often more applicable to more advanced WSPs. This can, though, be simplified by the fact that an operational limit commonly will control more than one hazard, and the most resistant or difficult to remove hazards (chemical and microbiology) can be used as a good surrogate for a range of parameters.

5.9.6 Water and chemicals traceability

When developing a treatment WSP it can be helpful to implement so called "*chemical and water traceability schemes*". Such schemes are powerful tools to assist in WSP management and decision making. Traceability helps the operators to identify, within an error margin, at what point in the plant is the water treated on that day, what were the operational conditions applied, and which water quality results were obtained. This type of information will be helpful in many situations. For example, using a traceability scheme the water utility will be able to:

- Develop objective cause analysis when deviation occurs, since it is possible to correlate the moment of the deviation with the different stages of treatment and operational conditions applied, thus enabling identification of potential events which might have caused it.
- When a deviation occurs, for example a critical limit exceedance, it is possible to identify where any potentially untreated water might be.
- With the traceability scheme, if a problem is identified with any chemical it will be possible to know if the chemical has already been used, when it was used and where the water treated with that chemical is now within the distribution system.

Developing a traceability system is an easy task that can be EXCEL based, or even set on a group of tables. The traceability should consider the minimum and maximum residence time in the treatment units, to allow a margin of error associated with uncertainty (Sancho et al 2008).

5.10 TREATED WATER QUALITY TARGETS

5.10.1 Introduction

The objective of water treatment is to provide water at the point of use which is safe and acceptable to consumers, although the actual water quality goals will legitimately vary between countries and locations (see Chapter 1). A WSP is the primary means by which this objective is secured and the water treatment WSP is a particularly critical part of the overall catchment to consumer approach.

However, it is known that even if water quality leaving the treatment works meets all necessary requirements, this can deteriorate in respect of some parameters during passage through the distribution network and also within consumer's premises. Thus, to secure all the defined water quality goals at consumer's premises, it is usually necessary to set more stringent operational treatment targets for certain key parameters in water leaving the treatment plant.

5.10.2 Water quality operational targets ex waterworks

The operational targets for water leaving the treatment works necessary to secure water quality goals at point of use, will depend on many different factors. This could include the length and condition of the network, the extent to which different source waters are mixed within the network, and the types of consumer premises and plumbing systems served. Full details of the water quality hazards and risks within the distribution network are given in Chapter 6 and for consumer's premises in Chapter 7. However, Table 5.3 below summarises some of the parameters which need to be considered:

Table 5.3 Parameters that could be used for ex waterworks operational targets

Parameter group	Potential water quality impact
Turbidity, organic nutrient (TOC, AOC), microbial parameters.	Microbial regrowth, tastes, odours, invertebrates, risk of increased disinfection byproducts, depending on residual disinfectant and water age.
Metal residual from raw water or coagulant, turbidity, organic material.	Discolouration deposits and sediments on pipe surfaces.
pH, alkalinity, corrosion inhibitor e.g. phosphate (if used).	Corrosion of distribution pipes and plumbing systems depending on materials used.
Inorganic parameters.	Normally conservative and thus unlikely to change much in distribution. However, blending with other source waters may need to be taken in to account for regulatory parameters such as nitrate. Note that the mineral composition of some waters treated through ion exchange, desalination or membranes may have increased metal corrosivity in some circumstances.

5.11 SITE SECURITY FOR WATER TREATMENT PLANTS

Chapter 3.9 describes a number of generic risks that can impact across all aspects of a water utility's operations, including water quality. Assessing and mitigating such risks may already be covered by the utility's overall risk management strategy, but this also needs to be linked specifically to the development of a WSP. Ensuring water treatment plant site security against vandalism or malicious intruders is a particularly important task because of its critical position in the whole supply system and the severe public health consequences if water quality is compromised through unauthorised access to sensitive plant or controls (Hrudey and Hrudey 2004).

However, site security should not simply protect against malicious activity, but also ensure that particularly vulnerable parts of the plant can only be accessed by those that need to do so for specific operational purposes. Of necessity, water treatment plants often have a significant number of workers, contractors and other visitors on site at various times. Thus site security has to cover both:

- Protection of installations, equipment and water from malicious actions.
- Control of access to critical installations and equipment to avoid inappropriate activities.

The degree of security necessary can only be assessed locally, taking into account a range of factors, including the sensitivity of each site operationally and in terms of vulnerability to unauthorised access. Different levels of security will also be necessary for those plants which are continually manned and those which are unmanned, but Table 5.4 below indicates some approaches that might be considered.

Table 5.4 Typical site security options for water treatment plants

Location	Security measure
Treatment assets	The plant perimeter should be protected by suitable fencing and lockable gates. Small unmanned sites (e.g. groundwater plants) should be contained within sealed and locked buildings. In vulnerable locations or at unmanned sites, additional security measures, such as use of closed circuit TV or intruder alarms, may need to be considered. All particularly sensitive locations within the plant, including restricted areas defined under a hygiene code (Chapter 3.5.5), should be separately locked, with access only allowed to authorised persons. Appropriate signage can be used to confirm whether, for example, specific permits to work are required in that area.
Procedures	Strict procedures should cover access to the site by staff, external contractors, or other visitors. Strict procedures should be in place to identify who can access which part of the site. Regular inspection of the site should take place by internal or contract security staff. Where appropriate, unmanned sites may also be patrolled by police or other civil authorities. Emergency procedures should include measures to deal with unauthorised access to sites.

REFERENCES

Bouchier et al (1998) Cryptosporidium in Water Supplies: Third Report of the Group of Experts to UK Department of the Environment, Transport and the Regions & Department of Health. Available from http://dwi.defra.gov.uk/research/bouchier/title.pdf

DWI (UK Drinking Water Inspectorate) (2010) Guidance on the use of UV irradiation for the disinfection of public water supplies.

Everall, N.C. and Lees, D.R. (1996) The use of barley straw to control general and blue-green algal growth in a Derbyshire reservoir. *Water Research* **30** (2), 269–276.

Everall, N.C. and Lees, D.R. (1997) The identification and significance of chemicals released from decomposing barley straw during reservoir algal control. *Water Research* **31** (3), 614–620.

Fewtrell, L., Bartram, J. (2001) Water Quality: Guidelines, Standards and Health – Assessment of Risk and Risk Management for Water-Related Infectious Disease. Published on behalf of WHO by IWA Publishing.

Foundation for Water Research (2010) Report FR/R0007 Causes of Copper Corrosion in Plumbing Systems.

Garbett R. (2005) An investigation into the application of floating reed bed and barley straw techniques for the remediation of eutrophic waters. *Water and Environment Journal* **19**, 174–180.

Hayes, C.R, (2010) Best Practice Guide on the Control of Lead in Drinking Water, IWA Publishing, London.

Health Canada (2002) Canadian Federal-Provincial-Territorial Committee on Drinking Water and the Canadian Council of Ministers of the Environment Water Quality Task Group: From Source to Tap - The multi-barrier approach to safe drinking water.

Hendry, K, Sambrook, H., Underwood, C., Waterfall, R., Williams A., (2006) *Water and Environment Journal* **20** (3), 159–168, Eutrophication of Tamar Lakes (1975–2003): A case study of land-use impacts, potential solutions and fundamental issues for the Water Framework Directive.

Heringa, M.B., Harmsen, D.J.H, Beerendonk E.F., Reus A.A., Krul C.A.M., Metz D.H., Ijpelaar G.F., (2011) *Water Research* **45** 366–374, Formation and removal of genotoxic activity during UV/H2O2eGAC treatment of drinking water.

Hrudey, S., Hrudey J., (2004) Safe Drinking Water: Lessons from Recent Outbreaks in Affluent Nations. Published by IWA Publishing.

Le Chevallier, M.W., and Au K-K (2004) Water Treatment and Pathogen Control: Process Efficiency in Achieving Safe Drinking Water. Published by IWA Publishing, London.

Murray, D., Jefferson, B., Jarvis, P. and Parsons, S.A. (2010) Inhibition of three algae species using chemicals released from barley straw. *Environmental Technology* **31**, 455–466.

Rosa, M., Paula, V., Menaia, J., (2009) O Tratamento de Água para Consumo Humano Face à Qualidade da Água de Origem. Published by Instituto Regulador de Águas e Residuos. Série Guias Técnicos, 13.

Sancho, R., Calvinho, A., Fontinha, P., Costa, L., Coelho, R., Lucas, H., (2008) Implementação de um Plano de Segurança da Água, Certificação de Acordo com a ISO 22000 e Certificação do Produto Água para Consumo Humano: Caso Prático. Published by Associação Portuguesa de Engenharia Sanitária e Ambiental, Maio a Agosto 2008, Série III, nº7, 28–35.

Sancho, R., Calvinho, A., Fontinha, P., Costa, L., Coelho, R., Lucas, H., (2009) IWA Drinking Water Quality International, **1**(3), Drinking Water Quality and Safety Management Systems – Águas do Algarve, SA experience in Portugal.

Van der Kooij, D. (2003) Managing regrowth in drinking water distribution systems, in Bartram et al., Heterotrophic plate counts and drinking-water safety: The significance of HPCs for water quality and the human health. Published on behalf of WHO by IWA Publishing.

Water Research Foundation (2010) Report 4088- Impact of chloride/sulphate mass ratio (CSMR) change on lead leaching in potable water.

WHO (2011) Guidelines for Drinking Water Quality, 4th Edition, World Health Organisation, Geneva.

WHO (2011b) Safe drinking water from desalination, World Health Organisation, Geneva.

Chapter 6

Developing a distribution water safety plan

Bob Breach and Chris Rockey

6.1 INTRODUCTION

Whilst great efforts are made to treat raw water to ensure it meets necessary quality standards, they can be seriously undermined by contamination or other forms of water quality deterioration in the distribution network. A drinking water quality management plan that is solely or mainly focused on treatment processes, and neglects preventative actions within the distribution network, runs the risk of operational inefficiencies and, potentially, the delivery of unsafe or unacceptable drinking water.

The proportion of the global population that receives piped water directly into their home is over 50% (WHO-UNICEF 2010) and with increasing urbanisation and current development trends this figure will rise. Therefore, from a global perspective, managing drinking water quality in the distribution network will become an ever more critical component of delivering safe and wholesome drinking water.

In water quality terms, distribution networks are often much more difficult to manage than water treatment works. This is because they:

- Are mostly buried underground and geographically widely dispersed.
- Provide a complex and often poorly understood asset base, sometimes in poor condition.
- Effectively become a large chemical and microbiological reactor with a large pipe surface to water volume ratio and wide range of potential interactions.
- Require significant costs to develop, maintain and repair.
- Generally have fewer barriers against external contamination than at a water treatment works and are more easily accessed by third parties.

It is for these reasons that water suppliers can find the development of a WSP for networks more difficult than one for a water treatment works. Nevertheless, research has found that, even in highly developed countries, networks can be a source of significant water quality problems. In developing countries the level of water quality risks may be even higher. For example, the percentage of water borne disease outbreaks attributable to networks was (Ainsworth 2004):

- 22% in the USA (1991–1996)
- 36% in the UK (1911–1995)
- 20% in Scandinavia (1975–1991)

In addition, it is widely known that distribution networks can be a significant source of consumer dissatisfaction related to aspects of water acceptability such as discolouration and taste.

Due to the problems identified above, when developing a distribution WSP, it is important to recognise that this can only be truly effective if, in parallel, there are close links with other utility management initiatives, particularly:

- Asset management - to understand the condition and performance of the network and develop appropriate short and long term maintenance programmes.
- Network hydraulic management - which is intrinsically linked to water quality performance.
- Finance and charging - since maintenance of the network will require significant resources.
- Consumers themselves - since they will need to be persuaded that delivery of high quality water does require significant long term investment and ongoing maintenance of the network.

Additionally, where the responsibility for local distribution of water to consumers is separate from water production and treatment (e.g. bulk supplies of treated water to a local distribution utility), then the relevant utilities must liaise carefully and have their water quality management systems closely aligned.

There are a wide range of potential water quality risks that may need to be managed within the water distribution network, many of which have complex inter-relationships. It is often the case that many of the basic components of a distribution WSP are already in place to varying extents within most water suppliers. However, the level of knowledge and resources necessary to develop a comprehensive WSP mean that this can take some time to fully implement, particularly for a large and/or complex distribution network. The challenge is thus to properly evaluate all of the risks which might occur and introduce prioritised, cost effective mitigation measures over a suitable timescale and taking account of the necessary resources.

Measures to reduce typical water quality risks will broadly fall into a number of categories. As a priority, all necessary training and procedures for the WSP should be put in place. In the longer term a better understanding of the network, leading to a comprehensive long term network asset management plan, will be necessary. Specific mitigation measures might typically include:

- Effective operational procedures using well trained and competent staff.
- Improving maintenance and security at vulnerable points within the network, such as treated water storage facilities.
- Developing improved knowledge of network condition, and hydraulic/water quality performance.
- Proper network management to minimise the risk of flow surges and low/zero pressure events.
- Improvement of the quality of water entering the network from the water treatment works.
- Long term maintenance strategies to minimise pipework deterioration and accumulation of sediment/deposits.
- Effective management of microbial growth by a combination of minimising water stagnation and residual disinfectant management.

- Education and awareness raising among consumers and others on the risks from backsiphonage and cross connections.
- Effective communication with consumers to respond effectively when things go wrong.

A simple diagrammatic representation which covers some of the main water quality risks typically identified within a distribution network is shown in Figure 6.1.

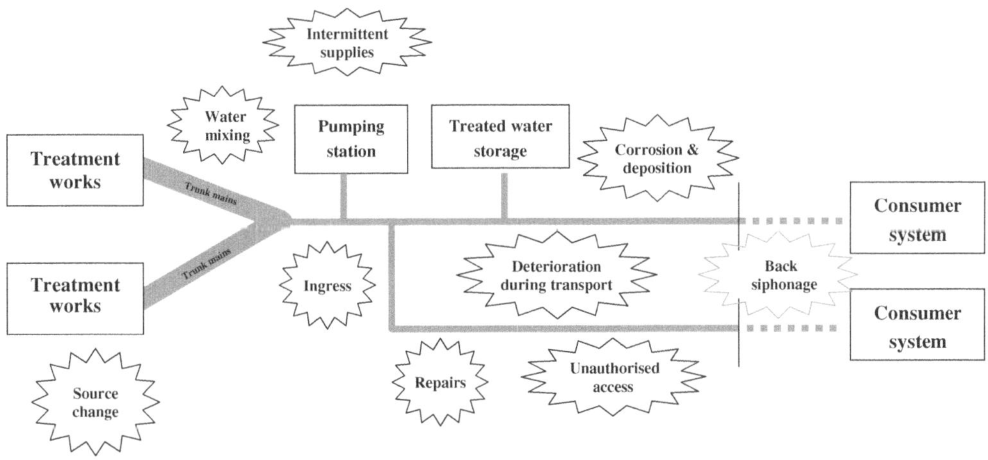

Figure 6.1 Water quality risks typically identified within a distribution network

These are described more fully in the following sections of this Chapter. Each describes; typical water quality risks which need to be considered based on an assessment of the likelihood and consequence of potential hazards and hazardous events; typical mitigation measures; and possible control points. Further technical information can be found in key references; for all risks (Davison et al 2005), for microbial risks (Ainsworth 2004), and discolouration risks (Vreeburg and Boxall 2007).

6.2 PUMPING STATIONS

Water is normally pumped into the distribution network from the water treatment works and then, where necessary, pressure is further boosted at additional pumping stations strategically located across the network. The design and location of pumping stations is primarily a hydraulic engineering issue, but there can be issues related to water quality which need to be taken into account as part of a WSP.

6.2.1 Water quality risks

The main water quality risks which can be associated with pumping stations are:

- Too high pressure: this can lead to enhanced air dissolution in the water which later outgases when leaving the consumer tap, thus giving rise to cloudy or "*milky*" water.
- Very variable pressure: this can lead to flow surges and thus disturbance of mains sediment which may be present.

- Pump failure or loss of power: depending on the design and configuration of the network, unless backed up by other pumps, this can lead to a number of water quality issues:
  - Loss of pressure leading to risk of contaminant ingress (see Section 6.9 below).
  - Loss of critical blending capacity, e.g. where high and low nitrate sources are being blended to secure regulatory compliance.
- Negative pressure created when pumps stop, leading to risk of contaminant ingress or other problems.

6.2.2 Risk mitigation measures and control points

Minimising the risks identified above is essentially a matter of:

- Proper hydraulic design in order to maintain a balanced and appropriate pressure across the network.
- Effective long term maintenance strategies.
- Incorporation of appropriate surge protection devices at pumping stations.
- Consideration of pump back up arrangements for critical pumping stations.
- Installation of suitable air (pressure relief) valves.

The best control point for this is network pressure measured at strategic points within the distribution system.

6.3 TREATED WATER STORAGE

Most distribution networks have some form of treated water storage to balance short term variation in demand. In some cases this also provides strategic storage to cope with bursts or other temporary disruption to the supply system. This storage can vary in size from small local steel tanks up to major concrete engineering structures with multiple compartments. Treated water storage structures are particularly at risk of contamination by ingress, since the water is not under pressure. Once contamination has occurred there may be little opportunity to rectify the problem before it reaches consumers.

6.3.1 Water quality risks

There are a number of risks associated with treated water storage reservoirs. These include:

- Microbiological contamination through ingress of contaminated surface water via faults in the roof/walls or through defective seals around access points or ducting.
- Malicious damage, potentially including deliberate or accidental introduction of a range of contaminants.
- Deterioration of water quality due to stagnation and poor turnover which can give rise to microbial regrowth and thus regulatory failures and/or tastes and odours.
- Interaction with construction materials causing tastes/odours or promotion of microbial regrowth.
- Deposition of debris which can subsequently become disturbed and enter the distribution network leading to discoloured water.

6.3.2 Risk mitigation measures

The risks identified above can be substantially mitigated by a comprehensive reservoir inspection and management plan which should form part of any overall asset management strategy. The type, location, condition and performance of each storage facility should be noted in an asset register and plans developed for both short and long term maintenance. Aspects impacting water quality which need to be considered by water suppliers, are set out below.

6.3.2.1 *Design and security aspects*

The detailed design of the structure will depend on a number of local factors. However, as far as possible, the following aspects should be taken into account.

- Treated water storage reservoirs should always be closed structures, designed to prevent contamination through ingress of surface or groundwater.
- The design should encourage turnover and circulation of the stored water within the structure.
- Secure facilities for access, cleaning and drain down. This could include consideration of how the structure can be taken out of service without impacting on consumers. For example, in larger reservoirs, by having separate multiple compartments.
- Materials of construction, including any membrane or sealant, should be approved for use with potable water.
- Hatch and vent covers should be secure and designed to prevent surface water ingress and access by birds, small animals, and insects.
- Water quality sampling facilities should be designed so that a representative sample can be obtained without inadvertently leading to contamination. Ideally, this should be via dedicated sampling arrangements, but if access covers have to be opened for dip sampling, then the design should minimise risks of contamination.
- In long distribution networks, storage reservoirs may be suitable locations for additional booster disinfection, provided this can be carried out with due regard to security and safety.
- Depending on the location, it will be necessary to include appropriate security arrangements. This could include animal proof fencing and secure locks for access hatches. In more sensitive or vulnerable locations, additional security measures might be necessary.

6.3.2.2 *Inspection and maintenance*

Treated water storage facilities should be included in routine inspection, sampling and maintenance programmes. The frequency and scale of each will need to take into account any regulatory requirements or national/regional codes of practice. It may also vary depending on local risk assessments with some larger, critical or higher risk sites having higher frequencies than others.

Storage facilities should be inspected externally on a fairly frequent basis to note any sign of damage, deterioration, evidence of animal or intruder activity or other factor which might prejudice water quality or other aspect of operational performance. Records should be maintained of each inspection and any necessary follow up action. Those carrying out the formal inspections should be appropriately trained. However, any operational staff visiting the site for other reasons should be encouraged to note and report any concerns immediately.

Less frequent internal inspections should also be programmed using a risk prioritisation approach or when evidence from external inspection or water quality monitoring indicates a potential problem. However, when doing so, particular care must be taken to ensure strict hygienic work practices using specially trained staff (see Chapter 3). After internal inspection, additional disinfection is normally recommended and, where possible, water quality checks carried out before returning the reservoir to service.

Routine maintenance programmes should be developed appropriate to the design, condition and operational criticality of the structure. Additional maintenance might be necessary as a result of external or internal inspections, or when water quality monitoring indicates a potential problem. The maintenance programme could include:

- Replacing or repairing faulty covers, vents and security equipment.
- Resealing faulty joints or roof membranes with appropriate materials.
- Resealing faulty cable or other access ducts.
- Cleaning accumulated debris from the internal floor of the structure.
- Cleaning and disinfection of walls and other surfaces.

6.3.3 Typical control points

Although the most suitable control points for water quality risks associated with treated water storage facilities can only be determined locally, there are likely to be two basic approaches, as set out below:

6.3.3.1 Sampling and analysis

Regular operational sampling of water leaving, or within storage facilities, provides useful trend information to monitor the onset of potential water quality problems. Frequent sampling, with analysis for a small number of key parameters, is probably more useful than less frequent sampling for a larger number of parameters. Key determinands could include simple microbiological tests (*E. coli*, coliforms, heterotrophic plate count) and chemical tests (chlorine, turbidity). As previously described, sampling arrangements must be designed appropriately in order to be representative of stored water, whilst at the same time not leading to inadvertent contamination, for example when dip sampling from open hatches.

6.3.3.2 Asset integrity

Routine frequent external checks on the physical integrity of the structure, for example hatch covers and security fencing, can provide important control points for water quality and are simple to implement. Routine, but less frequent, internal inspections of the condition and physical integrity of the structure also provides a very useful control.

6.4 WATER MIXING OR CHANGE OF SOURCE FEEDING NETWORK

Mixing of treated waters from different sources within the distribution network has been common practice for many years in a number of countries. This can be for a number of operational reasons, including enhancing reliability of supplies, helping to maintain regulatory compliance by mixing water of different chemical composition, or providing back up if a treatment works has to be taken out of use. It is also sometimes necessary to change the source of supply to a distribution network. Although mixing of waters or change of source can have a number of benefits, if this is not carried out with care there are risks that water quality in the distribution network may deteriorate. These are covered in more detail later, but are also reviewed briefly below.

6.4.1 Water quality risks and their mitigation

There are a number of potential water quality risks associated with mixing waters from different sources, most of which can be avoided by good planning before any changes are made. As part of any WSP it is important to fully describe the water supply system from source to tap. This should include a detailed assessment of the different characteristics of each water source before it is introduced, or mixed with other sources in the network. This assessment should include not only the natural mineral and organic constituents, but also the extent to which these might vary during the year, including both the chemistry and also physical factors such as temperature. It is also important to review any risks associated with the raw water catchment, and the capacity of the treatment plant to deal with any unexpected raw water pollution. Finally, it is necessary to have a good understanding of the network into which the supplies will be mixed or changed. Issues to consider include the following:

6.4.1.1 Effects due to hydraulic changes

Whenever water from different sources is mixed within distribution pipes, there is the potential for changes in water pressure, velocity and direction. This can affect both the rate of deposition of material on pipe surfaces and sediment accumulation, and also, in some situations, lead to material/sediment re-suspension.

Hydraulic pressure changes within a network could also lead to increased risk of contaminant ingress. By contrast, creation of high pressure within the network can enhance the dissolution of air in the water, which can then result in complaints from consumers of so called "*milky*" water.

6.4.1.2 Changes to the nature of mains deposits

Changes to the chemical composition of waters entering a network can have an impact on the rate and extent to which new deposits on pipe surfaces are formed, or can alter the nature of existing deposited material. This can be a particular problem, for example, when waters from very different sources are mixed, such as soft upland water with hard groundwater, or river water with groundwater. In addition, corrosion products within distribution networks that have historically been served by a relatively hard groundwater source, can be destabilised if that water is replaced with much softer water, for example from an upland reservoir. This can in turn lead to major discolouration problems, as the existing deposits become softer and therefore more likely to become disturbed by velocity changes.

6.4.1.3 Corrosion

Water is naturally corrosive to many metals and other substances, but the corrosion potential can be significantly influenced by the chemical composition of the water. This particularly relates to the relationship between pH and alkalinity, which gives rise to different carbonate equilibriums. Mixing of different waters can disturb such equilibriums, which in turn will have potentially significant effects on the dissolution potential for plumbing metals such as lead, copper and brass. Furthermore, if corrosion inhibitors such as phosphate or polyphosphate are used, then mixing of waters can impact on the extent to which these are effective within the network. By contrast, in some situations mixing of two chemically unstable waters can reduce overall corrosion potential (see Section 6.6).

An additional complication can arise when very soft water is used in areas previously supplied with hard water. In these circumstances, some concrete pipe linings can become affected as calcium leaches from the concrete matrix. In other areas, where pipes have historically been lined with coal-based or other tar material, different water types can influence the extent to which such pipe linings degrade and become unstable, thus leading to leaching or deposition within the body of water.

6.4.1.4 Organic and microbiological factors

Trace organic compounds in different water types can significantly affect the taste of water for consumers, particularly with surface water which does not undergo some form of advanced treatment such as activated carbon and/or ozone. Thus, introducing a surface water into a network previously supplied with only groundwater could lead to changes in taste and therefore consumer complaints, even though the water meets all the regulatory standards.

A critical aspect of water mixing is control of biofilm regrowth and the associated level of network disinfection residual that is required. It is well known that most surface waters have a higher natural organic content, which can stimulate enhanced biofilm growth on the internal mains surface. If not properly controlled, this biofilm can give rise to a number of microbial and taste problems at consumer taps. Maintaining microbiological safety requires a careful balance between disinfection barriers at the waterworks and, where necessary, the maintenance of suitable disinfectant residuals within the network. Maintaining such residuals at levels which effectively control microbial regrowth, without giving rise to consumer taste complaints, requires careful control. This is made much more difficult when water is being mixed from a number of different sources.

6.4.1.5 Other consumer related issues

As well as impact on normal regulatory parameters and those of concern to consumers such as taste, odour and discolouration, other factors may also need to be assessed when mixing or changing source waters. This particularly relates to any significant change in the overall mineral composition, and especially hardness, which is very noticeable to consumers. Additionally, for certain industrial processes which rely on water from the distribution network, many components of water can also have a major impact on production or manufacturing processes. Water suppliers do not normally guarantee any particular mineral composition of their water supplies, but may wish to provide key industrial water users with information on ranges of water quality which will normally be supplied (see Chapter 7). They may also have procedures for warning their consumers when these are changed.

6.4.2 Typical control points

The most effective control points for the risks described above are primarily associated with management procedures. All changes to the network which involve introduction of new sources of supply, or mixing water from different waterworks, should be carefully assessed by suitably experienced staff to identify whether any potential adverse water quality effects could occur. Appropriate mitigation measures should then be adopted before any change takes place. If a change in water composition has to be introduced which may be noticeable to consumers, for example a different mineral composition, then it will be important to have properly designed consumer communication programmes before the change is made.

6.5 CORROSION AND MAINS SEDIMENTS

Over time, many distribution networks accumulate sediments and other material deposited on the pipe surfaces. However, the nature of this material and the rate of deposition can vary considerably and the causative factors can be complex. Depending on the nature of the deposited material, it can be re-suspended by shear and other forces arising from change in flow velocity due to planned or unplanned activity such as bursts, increased demand or starting of booster pumps. This can give rise to, sometimes regular, discolouration events at consumers' taps. This is unacceptable to consumers and can damage their overall trust in water quality.

6.5.1 Water quality risks

The causative factors leading to sediment deposition can be varied and complex (Boxall and Saul 2005, Husband and Boxall 2010), but typically they arise through a combination of one or more of the following, as illustrated diagrammatically in Figure 6.2:

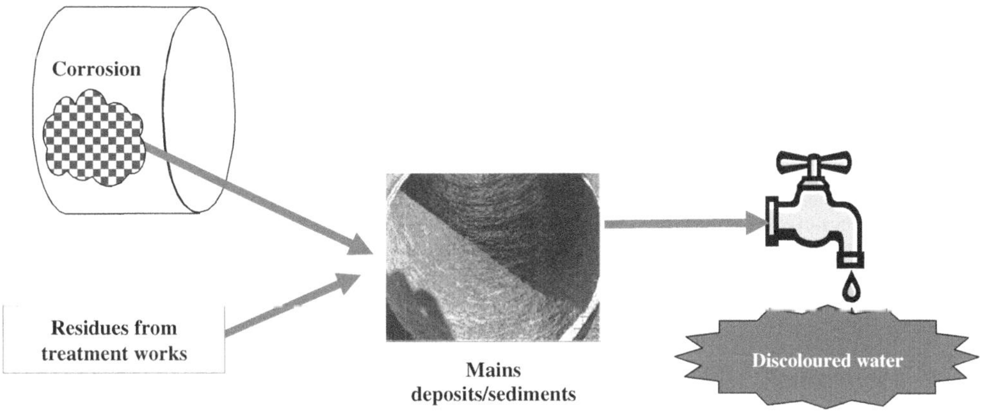

Figure 6.2 A simple summary of discolouration processes in networks

6.5.1.1 Residues from treatment works

A variety of metal cations originate from treatment works. These can include iron or aluminium coagulants, as well as naturally occurring iron, aluminium and manganese. Even a well managed large surface water treatment works with iron residuals of, say, 0.05 mg/l will be putting several tonnes of iron a year into the network. In addition, in some situations, natural humic organic matter can co-precipitate with cations in the network to form complex organic/inorganic deposits.

6.5.1.2 Corrosion products

Iron oxides and other debris from corrosion of iron pipes can accumulate over time which, as well as impacting on water quality, can adversely affect flow and pressure.

6.5.1.3 Biofilm

In some cases, where biofilm growths are significant, they can bind metal residues, thus forming sediment comprising complex, discoloured, biological matrices. Such biofilms can also facilitate the creation of localised areas on the pipe surface which have accelerated corrosion potential.

6.5.2 Risk mitigation measures

Mitigation measures to minimise sediment/deposit formation, and thus the likelihood of discoloured water, will depend very much on local circumstances. Many of the measures require significant resources, but, since sediment formation is typically a long term problem, the measures set out below should be part of an asset management plan, phased over an appropriate timescale, with priority given to those measures that:

- Deal with areas where the risk of discoloured water is highest.
- And/or will have the most cost effective long term outcome.

6.5.2.1 Waterworks residue management

Ex works residual levels of materials, including iron, aluminium or manganese, as well as overall turbidity and levels of natural organic matter, should be maintained at the lowest economically practicable level. This can be achieved through better optimisation of existing treatment and/or treatment asset improvement.

6.5.2.2 Mains cleaning/flushing programmes

There are a number of different techniques for this, but all incur some loss of water so need to be managed carefully, particularly in times of, or areas prone to, drought. Careful planning, including consumer pre-warning, will be necessary to minimise impact on consumers.

6.5.2.3 Network design

As far as practicable, networks should be designed and operated to maintain the lowest possible age of water and also minimise the risk of major flow surges and shear forces. This can be achieved, for example, by minimising long mains with low demand, avoiding dead ends in the system, and managing booster pump arrangements.

6.5.2.4 Network rehabilitation programme

A long term mains rehabilitation/renewal programme should be developed to renovate areas of the network suffering from unacceptable levels of mains corrosion, and replace/reline unlined iron pipes as part of an overall asset management plan.

6.5.2.5 Managing water mixing

Where waters from different source types are mixed in the network, chemical reactions can increase the formation of new sediments, or alter the nature of existing sediment (see Section 6.4).

6.5.2.6 Biofilm management

In some situations high levels of biofilm can also influence the level and type of sediments (see Section 6.7).

6.5.3 Typical control points

The most effective control points will depend on an assessment of the most likely causes of sediment/deposits in the network. This can be helped greatly by determining the extent and nature of existing sediments/deposits, through evaluating such things as:

- Number of consumer complaints of discoloured water.
- Chemical/microscopic analysis of sediments to determine the potential sources.
- Visual inspection of the internal levels of sediment/deposits and extent of any corrosion, when mains are opened for repair or other reason.

Analysing these data and plotting them spatially using GIS, or creating network models, can greatly enhance understanding of the main underlying causes and the areas of highest risk of discolouration. Based on this assessment, control points could then include:

- Concentrations of key parameters leaving treatment works e.g. iron, aluminium, manganese, turbidity.
- Regularity of agreed mains flushing programmes.

- Maintaining appropriate pressures and level of flow variability in the network.
- Modelling to ensure that a predefined maximum water age is obtained across the network.

6.6 MAINTAINING EFFECTIVE CORROSION CONTROL

Water can be naturally scale forming, but can also be corrosive to many metals and other substances. The corrosion potential can be significantly influenced by the chemical composition of the water. It is, therefore, common to control the corrosion potential of water leaving the water treatment works through various types of "*water conditioning*", particularly correcting alkalinity/pH, or in some cases adding corrosion inhibitors (see Chapter 5.8). This is designed to protect both the network itself and also consumer plumbing and fittings. However, there is a risk that the effectiveness and level of corrosion control can change as water passes through the network.

6.6.1 Water quality risks

The primary factor impacting on the corrosion potential of water is normally the relationship between pH and alkalinity, which gives rise to different carbonate equilibriums. Ideally, water should have a neutral or slightly positive saturation index with respect to calcium carbonate. However, this can change during passage through the network, for example because of interaction with cement based pipe materials or through change in dissolved CO_2 levels. The carbonate equilibrium can also be affected by mixing of different waters within the network. Any change in carbonate equilibrium could, in turn, have potentially significant effects on the dissolution potential in relation to plumbing metals such as lead, copper and brass.

Some water suppliers add corrosion inhibitors, such as phosphate or polyphosphate, to further reduce the corrosion potential of the water to plumbing metals, although this is often covered by local regulations. Where such inhibitors are used, it is important to assess the extent to which their level can change by mixing or dilution as water passes through the network and through any unexpected interactions (Hayes 2010).

Although it is the carbonate equilibrium which particularly affects corrosion potential, other aspects of the mineral composition can also be important. This includes, for example, the chloride to sulphate ratio (Water Research Foundation 2010) and, in some situations, dissolved oxygen levels.

6.6.2 Risk mitigation measures and typical control points

Risk mitigation primarily relies on an assessment of the likelihood of change in water corrosion potential as it passes through the network. This, in turn, can be greatly assisted by a good knowledge of the network and its condition, as well as typical plumbing materials used in consumer premises. This can be facilitated by well designed network models. Mitigation options include:

- Enhanced water conditioning at the waterworks, for example by improving the alkalinity/buffering capacity.
- Careful design and operation of networks to minimise water age and inappropriate mixing of waters.
- Choice of pipe materials appropriate to the water being transported.

Typical control points mainly relate to measurement of key water corrosion indicators such as pH, or concentration of corrosion inhibitor, at strategic points within the network.

6.7 MICROBIAL REGROWTH AND NETWORK DISINFECTION

6.7.1 Water quality risks

Water leaving the water treatment works should be completely free of harmful microorganisms and should also fully meet all relevant microbial standards including those for indicator organisms such as *E.coli*, and coliforms. Nevertheless, however well treated, the final water will always contain some residual heterotrophic organisms (HPC) as well as organic nutrient. Particularly in surface derived systems, these organisms may therefore regrow as water passes through the network and may also colonise the internal surfaces of water mains and other structures. Although not necessarily a health risk, this can lead to other water quality risks (Bartram et al 2003) and is thus often controlled by the use of residual disinfectants. However, this in turn can result in the formation of a wide range of disinfection byproducts, which are often covered by regulatory standards and can lead to unacceptable taste/odours to consumers.

The interaction between microorganisms, nutrient and chemical disinfectants is complex and also affected by a range of other factors, including temperature and time of travel though the network (Figure 6.3 below). The situation is amplified by the formation of biofilms on pipe and other surfaces. Typically, even in a "*biologically clean*" network there can be 10^4 bacteria/cm^2 of pipe wall, and this can increase by 2–3 logs in "*biologically dirty*" systems. It is thus common to use disinfectant residuals to manage regrowth within the network, as well as provide some limited protection against ingress of low level microbial contamination.

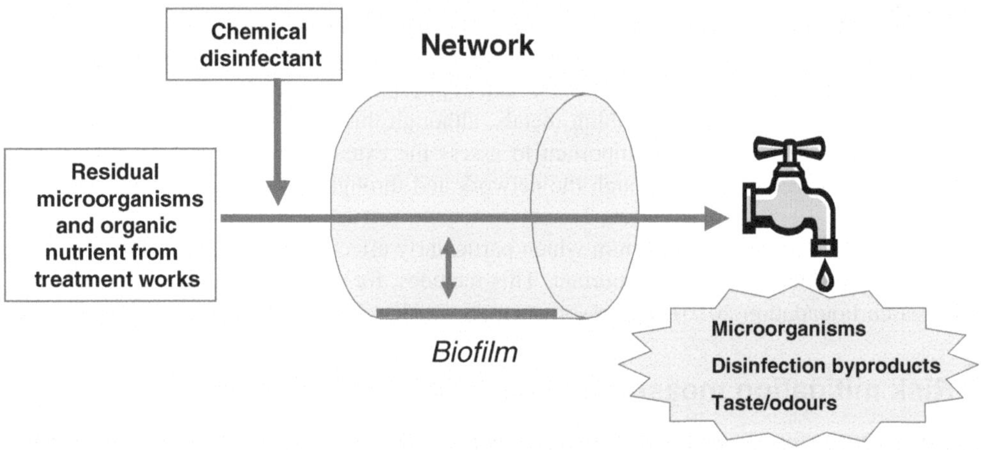

Figure 6.3 A simple illustration of interactions between microorganisms, nutrient and chemical disinfectants

Whilst it is essential to avoid any risk from pathogenic microorganisms, it is neither practicable nor necessary to have networks free of all microorganisms. The World Health Organisation (WHO 2011) advises that there is no evidence, either from epidemiological studies or from correlation with occurrence of waterborne pathogens, that heterotrophic organisms (HPC) alone directly relate to health risk. However, sudden increases in HPC levels might sometimes be associated with faecal contamination through ingress or other means and therefore need further investigation.

However, as well as numerous free-living heterotrophic bacteria, biofilms can also contain many other undesirable microorganisms. These include fungi, protozoa, nematodes and crustaceans, and may be an issue particularly in older systems which contain deposits and sediments formed by the internal corrosion of metal pipes and/or accumulation of treatment residues. This can give rise to a number of water quality risks which can increase in proportion to the level of biofilm. Problems can include:

- Adverse taste and odour of water delivered to consumers.
- Growth of aquatic organisms such as *Asellus aquaticus* which are highly unacceptable to consumers.
- Discoloured water associated with disturbance of iron stained biofilm.
- Potential "*masking*" of opportunistic pathogens or other undesirable bacteria.

Whilst it is very common to maintain disinfectant residuals within the network for microbiological control, this can, at the same time, also increase other water quality risks, especially disinfection byproducts (DBPs) and unacceptable tastes/odours for consumers. The levels of both DBPs and unacceptable tastes/odours are linked to the extent of causative factors such as natural organic matter precursors, scale of biofilm, water age and temperature. However, it should be noted that consumer acceptability of tastes is often influenced by both the absolute level and the variability in chlorine levels.

The standards which apply to DBPs vary between countries, as does the relative degree to which consumers will find chlorinous taste/odours acceptable. Thus the level of control necessary can only be determined locally. It is important to stress that the advice from the World Health Organisation (WHO 2011) is that microbial safety should never be compromised in an attempt to reduce DBP levels.

6.7.2 Risk mitigation measures

Risk mitigation strategies for controlling microbial growths in the network, whilst at the same time meeting locally determined requirements for DBPs and consumer taste/odours, can be complex, since they can be affected by a wide range of local factors. Risk mitigation should therefore be part of an overall asset management strategy. Potential measures, either on their own, or in combination, could include:

- Upgrading and/or improved operation at the waterworks. This can reduce seeding of microorganisms into the network (which may be protected in particulate matter) and reduce organic matter in the final water that can act both as nutrient for microbial growth and precursors for DBPs.
- Improved management of the network to reduce water age and areas of stagnation, including treated water storage facilities, and also avoiding velocity surges which can disrupt biofilms. In some cases, network models can help to target such improvements.
- Careful management of disinfection residuals within the network to keep them as low and stable as possible, consistent with good microbial quality. Depending on the network, this may be achieved solely through management of the residual levels leaving the waterworks. However, in some cases, the use of supplementary booster disinfection at strategic locations within the network can reduce the need for high residuals leaving the works and allow much more careful overall control of residuals. Some network models can also help with design of booster disinfection strategies.
- Choice of residual disinfectant. This is very much a matter for local evaluation and may also be impacted by local laws which could stipulate which disinfectant can, or cannot, be used (see Chapter 5).
- Introduction or review of existing network cleaning strategies to improve removal of sediments and other material which can exacerbate biofilms and other nuisance organisms.
- Review of the long term network rehabilitation strategy to increase the rate of renewal of mains in poor condition.

6.7.3 Typical control points

The most effective control points will depend on an assessment of the most likely risks and the potential causative factors. This can be helped by mapping the extent and nature of potential problems within the network using GIS or network models. Based on this assessment, control points could include:

- Concentration of key parameters leaving treatment works, e.g. disinfectant residual, DBPs, turbidity, organic nutrient.
- Concentration of microbial indicators such as *E.coli*, coliforms and HPC either ex works or at strategic points within the network.
- Concentration of disinfectant residual and its variability in the network.
- Concentration of relevant DBPs within the network.
- Numbers of consumer taste/odour complaints.

6.8 INVERTEBRATE INFESTATION

Invertebrate animals are naturally present in many raw water resources used for the supply of drinking water. Small numbers of adults, or their larvae, may pass through water treatment works, if the barriers to particulate matter are not completely effective. Many of these animals can survive, and some may even reproduce within the supply network, by deriving their food from the microorganisms and organic matter present in deposits on pipe and tank surfaces. Populations of small animals can be widespread in treated water distribution systems. However, the density and composition of animal populations vary widely, from heavy infestations of readily visible species that are objectionable to consumers, to sparse occurrences of microscopic species.

6.8.1 Water quality risks

For control purposes, the types of animal potentially found in distribution networks are divided by the World Health Organisation (WHO 2011) into two groups:

- Free-swimming organisms in the water itself or on water surfaces, such as the crustaceans *Gammarus pulex* (freshwater shrimp), *Crangonyx pseudogracilis*, *Cyclops* spp. and *Chydorus sphaericus.*
- Animals that either move along surfaces or are anchored to them (e.g. water louse *Asellus aquaticus*, snails, zebra mussel *Dreissena polymorpha*, other bivalve molluscs and the bryozoan *Plumatella* sp.), or inhabit slimes (e.g. *Nais* sp., nematodes and the larvae of chironomids).

Additionally, larvae of many species of chironomid may be present in the distribution system, on occasions in appreciable numbers. The main water quality issue from all such organisms is their unacceptability to consumers when discharged through taps, either directly or in association with discoloured water.

In tropical and subtropical countries, there are species of aquatic animal that act as secondary hosts for parasites, e.g. the small crustacean *Cyclops* is the intermediate host of the guinea worm. However, the World Health Organisation (WHO 2011) advises that there is no evidence that guinea worm transmission occurs from piped drinking water supplies. It has also been suggested that some invertebrates may be of sanitary significance, since they might "*mask*" potentially pathogenic bacteria from disinfection. However, again WHO advises that there is no evidence of any association of these organisms with waterborne gastrointestinal infection for the population at large.

6.8.2 Risk mitigation measures

The most likely cause of invertebrate infestation of the distribution network is penetration of aquatic animals through surface treatment processes, which then form breeding populations within the distribution system. In addition, invertebrates can enter through defects in the integrity of the distribution system.

The risk of infestation is more likely in parts of the network which are in poor condition, or where other factors favour invertebrate growth and reproduction, such as a good supply of food, warm water and slow water movement. Risk mitigation measures to prevent infestation occurring therefore fall into a number of categories:

6.8.2.1 At surface waterworks

Effective removal of particulate and organic matter at the waterworks will minimise the passage of invertebrates into the network and, particularly importantly, will remove potential organic food sources such as algae which can sustain the growth of invertebrates within the network. Additionally, it is important to ensure effective management of filters and filter media, for example through adequate backwash and also prevention of colonisation of filters or open tanks with chironomid species.

6.8.2.2 At service reservoirs

Effective maintenance to prevent access of insects, for example through uncovered vents or damaged hatch covers.

6.8.2.3 Within the network

It is important to ensure that routine management and maintenance of the network helps to prevent the creation of habitats favourable to invertebrate growth and reproduction. Such approaches might include minimisation of mains corrosion and mains deposits and minimising dead ends and low flow areas.

Once infestation of the network has occurred to such an extent that remedial action is required, then various physical mains cleaning methods need to be used. This also has the advantage of removing the sediments that provide habitat and food supply for animals, as well as the animals themselves. The choice of cleaning method will depend on the species of animal present, their ease of removal and the numbers present. It will also be impacted by other local factors including availability of technical equipment and resources, the extent to which sufficient water is available for flushing, and other practical issues such as arrangements for disposing safely of waste water. In general, species that move freely in the water or on the surface of the pipe or deposits (e.g. *Cyclops*) are relatively easily removed by flushing, whereas, those that burrow in deposits (e.g. nematodes, chironomid larvae) require action that is more stringent. Species that cling to the pipe surface (e.g. *Asellus aquaticus*, aquatic gastropod snails) require dislodging before they can be flushed from the main.

Effective mains cleaning requires accurate mains records and normally involves several valving operations. The cleaning process should work systematically downstream, starting at the treatment works if practicable. To reduce re-colonisation of cleaned mains, no main that has been treated or is being treated should receive water from an untreated main.

In some extreme situations, chemical treatment may be necessary to supplement mains cleaning. The most important of the chemical agents are pyrethroids, which are effective against a range of arthropods, including chironomid larvae and *Asellus aquaticus*. However, use of such chemicals must be carried out only with specially approved products and may be subject to local legislation. If used, great care must be exercised to prevent impact on sensitive consumers and sensitive water use, such as domestic and commercial fish stocks, as well as environmental damage from waste water disposal.

6.8.3 Typical control points

Control points used for other water quality risks at treatment works and within the distribution network will also help to prevent problems due to invertebrates. Additionally, where justified, regular monitoring of populations of animals in the distribution system will provide information on their relative abundance in different parts of the system and on changes in their numbers. This allows control measures to be taken in a planned manner and at a suitable time, before numbers become high enough to cause major problems. The usual method of sampling animals in water pipes is to flush a standard volume of water from a hydrant at a controlled flow rate, and to capture particulate matter, including any animals that might be present, in a fine-meshed sampling net.

6.9 INGRESS OF CONTAMINANTS TO THE NETWORK

6.9.1 Water quality risks

Normally, water distribution networks are pressurised, which minimises the risk that contaminants can enter. However, the risk of ingress can increase significantly when water pressure is low or zero. This can occur for a number of reasons, including planned repairs to the network, bursts in the water main, and excessive local demand on the system, for example during fire fighting or drought. Also, although not considered good practice, some systems in less developed countries only provide supplies on an intermittent basis leading to cycles of pressurisation and depressurisation. Once contamination has occurred there may be little opportunity to rectify the problem before it reaches consumers. Ingress can occur from a number of sources including:

- Contaminated water leaking into underground chambers on the network, such as pressure relief valves, control valves, strategic meters, and fire hydrants.
- Contaminants from sources such as tip sites, old industrial land, and sewage disposal facilities, which could, under certain circumstances, enter water mains in poor condition when they experience low or zero pressure.

Other sources of contaminant ingress of particular concern, which are covered in more detail later include:

- Backsiphonage from unsatisfactory or poorly installed plumbing in consumers' premises.
- Unauthorised access either for domestic use or for industrial applications.
- Mains repairs.
- Intermittent supply of water.

6.9.2 Risk mitigation measures and typical control points

The risk of ingress from mains and fittings can be substantially mitigated by a comprehensive network asset management plan, which should form part of any overall asset management strategy. The type, location, and condition of all mains and mains fittings should be noted in an asset register and prioritised plans developed for both short and long term maintenance. It is particularly important that underground chambers containing valves and other fittings are designed and located to minimise the risk of flooding. As far as possible they should also be located in areas where the risk of being buried or affected by local sources of contamination risk, such as animal or industrial wastes, are minimised.

The main control points for such risks are likely to be based on:

- Regular inspection of vulnerable points in the distribution network to identify whether they have become damaged, buried or put at risk by nearby sources of contamination.
- Routine microbiological monitoring of the distribution network to assess whether contamination may potentially be occurring from this cause.
- Monitoring consumer complaints.

Risk mitigation from the other specific sources identified above, e.g. backsiphonage, is covered later in the relevant section.

6.10 DISTRIBUTION NETWORK REPAIR PROCEDURES

6.10.1 Water quality risks

When the water distribution system needs to be repaired or updated, there is a major risk of treated water in the distribution network becoming contaminated, unless strict working procedures are implemented. These risks relate particularly to microbiological contamination, but can also include customer acceptability.

6.10.2 Risk mitigation measures

Most risks can be minimised through the rigorous adoption of appropriate working procedures, although these may vary depending on local factors.

6.10.2.1 Training in general hygienic work practices

It is of vital importance that all staff involved in repairs to the network are fully trained in hygienic work practice, and aware of the water quality risks that can occur if proper precautions are not taken. Many water suppliers use external contractors to carry out mains repair work and it is of equal importance that the training of any contractor's staff used for this work is as rigorous as that for internal staff. Much of the necessary approach is simple common sense and staff should be encouraged to identify and report any potential hazards or hazardous events which are identified during repairs. Records should be kept of all staff and contractor hygiene training, which should be updated at appropriate intervals.

It is particularly important that high standards of personal hygiene are maintained for all staff working on network repairs. Defecation and urination close to the repair site should be prohibited, except when using any proper toilet facilities which are provided. Measures should be in place to ensure that any staff suspected of suffering from, or being a carrier of, water borne disease are not allowed to carry out repair work until free from infection (see also Chapter 3).

6.10.2.2 Staff and contractor supervision

All repair work should be supervised and/or checked from time to time by suitably qualified and experienced staff to ensure that the correct working procedures are followed. Supervisors should also check that all staff and contractors have received suitable training.

6.10.2.3 Planning the work

As far as possible, any work on the distribution network needs to be carefully planned in advance to minimise risks to water quality and other adverse consumer impact. This could include necessary valving operations, provision of equipment and materials, responsibility for work, identification of consumers potentially affected (particularly vulnerable consumer groups or premises), and notification to those consumers/premises (see Chapter 7). The consumer call centre also needs to be briefed, so they can effectively respond to any consumer enquiries that might arise as a result of the repair work.

In emergency situations, such pre-planning will not be possible, but procedures should still take the above factors into account, including making sure that consumer call centres are notified.

6.10.2.4 Consumer notification

Any pre-planned work on the distribution network should, as far as practicable, be notified in advance to affected consumers. Information should include expected time and duration of any planned work, the potential for loss of supply, the possibility for short term discolouration after repair and contact phone numbers for any queries or concerns.

6.10.2.5 Tools and equipment

It is important that measures are taken to ensure that all tools, equipment and material used for the repair work are kept visibly clean and free from contamination. These could include:

- Procedures should ensure that any tools or equipment used for restricted water supply purposes have not previously been used in contact with sewage or for sewage related activities.
- As far as practicable, tools and materials should be kept clean whilst being stored on site, and disinfected using a fresh chlorine solution before use.
- Pipes and related fittings should be capped and/or covered whilst being stored, to ensure that dirt or animals cannot enter before use. They should be inspected internally and, where necessary, cleaned before installation.

6.10.2.6 Procedures during repair

It is important that documented procedures for effective repairs which avoid water contamination are developed for each type of repair work and strictly enforced. The contamination risk will vary according to the type of repair. For example, mains which can be repaired by a "*split collar*" whist remaining under pressure, provide a lower risk than mains which are completely broken and thus need to be completely depressurised before repair. Procedures could include:

- *Water ingress*
  It is essential that any hole or trench dug to carry out the repair is kept as dry as possible. The depth of the excavation should be lower than the bottom of the pipe to ensure that dirty water does not enter the pipe. If necessary, a pump should be used to keep the trench dry.
- *Sewage contamination*
  If a broken sewer pipe, or other evidence of faecal contamination, is found within, or close to, the excavation then this must be reported immediately to the supervisor. Additional precautions must

be taken to ensure that this does not result in water contamination. The exact response will depend on the circumstances and can only be assessed locally. However, this could include additional disinfection within the pipe, bacteriological testing of the water, and possible precautionary advice to consumers to boil water until the contamination is cleared.

- *Oils and other liquids*
  Liquids often used during repairs, such as oil, diesel, solvents, lubricants or paint, can cause unacceptable tastes for consumers if they enter the main. If plastic pipes are being used then such materials can also permeate through the pipe into the water. Procedures must ensure that such liquids are stored properly on site and are not allowed to contaminate the main or immediate area of operation.

6.10.2.7 Procedures after repair

Once the repair has been completed and the trench backfilled, there will be a number of necessary follow up procedures. Depending on the circumstances, these could include:

- Recharging the main in ways designed to minimise disturbance of deposits, sediments or debris.
- Using additional disinfection in the area of the network affected by the repair.
- Notification to the consumer call centre.
- Completion of necessary records.

6.10.3 Typical control points

The main approach to controls is likely to be based on:

- Maintaining complete and up to date training records for both internal and contract staff, coupled with verification of this by supervisors on site. Some water suppliers use a staff/contractor certification scheme to improve controls.
- Taking check bacteriological samples on water from mains in the vicinity of all, or a proportion, of the repairs to check that procedures have not resulted in contamination.
- Regular inspection of repair sites by suitably qualified and experienced staff to ensure that the correct working procedures are being followed.

6.11 LAYING NEW DISTRIBUTION MAINS

6.11.1 Water quality risks

When a water distribution system needs to be constructed, extended or remediated, there is a major risk of treated water in the distribution network becoming contaminated unless strict procedures are implemented. Many of the water quality risks and mitigation measures relating to new mains are the same as those which apply to mains repairs (see section above). However, in addition, there are a number of water quality risks specifically associated with the design, procurement and laying of new mains and fittings. These risks relate particularly to microbiological contamination, but can also include consumer acceptability.

6.11.2 Risk mitigation measures

The water quality risks and appropriate mitigation measures will vary, but could include:

6.11.2.1 Planning the work

Any major work to construct, extend or remediate distribution networks will need considerable planning to minimise risks to water quality and other adverse consumer impact. If the work is likely to impact consumers, then these need to be identified (particularly vulnerable consumer groups or premises, see Chapter 7) and notified about the work in advance. Information should include the time and duration of any planned work, temporary alternative supply arrangements, and contact phone numbers for any queries or concerns. The consumer call centre also needs to be briefed so they can effectively respond to any consumer enquiries that might arise.

6.11.2.2 Ground conditions

In some older urban areas, soils might historically have been contaminated with industrial or other waste material. If pipes are laid in such soil, then there is a risk that contaminants could pass through pipes, particularly if made of plastic, Contamination may also occur through ingress after pipe bursts. All potentially affected soils through which new pipelines may pass should first be assessed for contamination risk. If a risk exists, where possible alternative routes should be identified which avoid the contaminated areas. If this is not possible, special precautions will need to be taken, such as removing contaminated soils before main laying, use of special pipe protection measures, use of alternative pipe materials, or laying pipes in trenches lined with inert material.

6.11.2.3 Choice of pipes and equipment

It is important that pipes, or any other material, that will be in contact with treated water do not promote microbial growth or leach any potentially harmful contaminants into the water (see Section 6.16).

6.11.2.4 Network design

There are a number of network design risk factors which can lead to a deterioration of microbiological or chemical quality. These have to be balanced against hydraulic, engineering and other local factors, but it is important that these are assessed and addressed as part of project design and implementation (see Section 6.15).

6.11.2.5 Commissioning of new mains

Once the mainlaying work has been completed, it is important that there are documented procedures which ensure that there will be no contamination of the water when the new mains are introduced into service. Best practice procedures may vary slightly between countries, but should include:

- Flushing and/or swabbing the main until clear.
- Rigorous disinfection of the new main. Ideally, this should maintain an initial free chlorine concentration typically around 20–25 mg/l for 16–24 hours or equivalent.
- Flushing the main to remove excess chlorinated water, taking care to avoid environmental damage.
- Recharging with normal mains water and standing for a further 16–24 hours.

- Taking samples along the main for bacteriological analysis, ideally on two successive days.
- Bringing the main into service within a defined time period if samples are free of indicator organisms such as coliforms, and of acceptable appearance and taste.

6.11.3 Typical control points

The main approach to controls are likely to be based on:

- Maintaining complete and up to date training records for both internal and contract staff, coupled with verification of this by supervisors on site. Some water suppliers use a staff/contractor certification scheme to improve controls.
- Regular inspection of construction sites by suitably qualified and experienced staff to ensure that the correct working procedures are being followed.
- Checks that pipes and pipe fittings meet specified quality criteria.
- Taking check bacteriological samples on water from all new or remediated mains before putting into service to ensure that there is no water contamination.

6.12 INTERMITTENT SUPPLIES

6.12.1 Water quality risks

Significant numbers of consumers in developing countries across the world only receive piped water supplies on an intermittent basis. The reasons for this vary, but particularly include severe water shortage, poor water infrastructure and inadequate financing/cost recovery. The cycle of regular pressurisation/depressurisation can cause significant water quality risks from ingress of contaminants, particularly if the sewerage infrastructure is also in poor condition (Tokajian and Hashwa 2003). There are a number of potentially significant risks from intermittent supplies. These include:

6.12.1.1 Ingress of contaminants

As described elsewhere in this Chapter, when water distribution networks become depressurised, there is a much enhanced risk of contaminants entering water mains. This risk is increased significantly if sources of contamination, such as sewerage infrastructure, waste disposal facilities and industrial premises are also in poor condition. Risks particularly relate to microbiological contamination and water borne disease, but other forms of chemical contamination could also occur.

6.12.1.2 Stagnation within the network

When the network is depressurised there will still be water left in the system for periods of perhaps 24 hours or more. The quality of this water will degrade through microbial and chemical reaction, causing sediments, poor taste and often increased risk of failing other health related water quality criteria. When the pressure is restored this stagnant water may flush into consumers' premises.

6.12.1.3 Contamination within households

Because of the intermittent supplies, consumers will develop a range of practical coping methods. These could include collecting water from standpipes/street water sellers, increased household water storage and/or digging of shallow wells. Particularly in poor households, the construction, condition and maintenance of these facilities could be very inadequate, thus leading to further significantly enhanced risk of water contamination.

6.12.2 Risk mitigation measures

This is a complex issue with a range of much broader policy implications. For the long term mitigation and management of water quality risks associated with intermittent supplies it is strongly recommended to shift to continuous (24/7) supply . Provision of a fully pumped continuous water supply will not only benefit water quality but also enhance social and other aspects. However, as set out above, this is likely to require potentially significant resources and possible institutional change. Such a shift will, therefore, normally need to be part of a broader discussion on water policy, financing and social issues, and will thus require strong commitment from relevant institutions, such as national/regional governments as well as senior utility managers.

However, in the short term, where intermittent supplies are prevalent, the water quality risks may be mitigated to some extent through measures identified elsewhere in this Chapter, particularly in sections dealing with:

- Ingress of contaminants to the network.
- Unauthorised access to the network.
- Backsiphonage into the network.
- Distribution network repair procedures.
- Microbial regrowth and network disinfection.

Management of private household supplies is outside the scope of this book, but much useful information can found in relevant WHO publications.

6.13 UNAUTHORISED ACCESS TO THE NETWORK

6.13.1 Water quality risks

Illegal or unauthorised access to the distribution network can arise from time to time in many water supply systems, but the problem can often be particularly acute in some developing countries. Reasons include vandalism or other malicious intent, use of the water for commercial activity and building, and use by those who do not have access to, or cannot afford, piped supplies. This poses a number of water quality risks, particularly through contaminant ingress, and can also create wider problems related to loss of revenue. The water quality risks can be significantly increased if the unauthorised access occurs in an area where major sources of contamination exist, for example in areas with poor sanitation arrangements. The risks can occur from both:

- Direct backsiphonage of a range of contaminants at the point of access, including faecal matter, industrial chemicals and other material.
- Indirectly, by reducing water pressure, thus increasing the risk of backsiphonage elsewhere in the local vicinity.

Also, if the volume of water taken is significant, then this can adversely affect the overall operation of the distribution network by changing flow rates and other hydraulic factors.

6.13.2 Risk mitigation measures

Dealing with unauthorised or illegal access can be difficult, but typically relies on two main approaches:

6.13.2.1 Security

Points at which access is typically obtained include valve chambers or fire hydrants. Access by unauthorised staff can be made more difficult by measures such as improved and more secure design of valve chambers, and lockable or special design fire hydrant covers. Where it is appropriate to allow access to the network by third parties, for example for fire authorities and approved builders, then this should be conditional on proper training and possibly certification. Additionally, such access should be conditional on use of suitable backflow prevention devices.

6.13.2.2 Education

More generally, in areas where unauthorised access is common, then a programme of local community education should be established to highlight the risks.

6.13.3 Typical control points

Typically, the two main control points are:

6.13.3.1 Inspection

The easiest control to implement is regular inspection of the network to identify points at which unauthorised access regularly occurs, so that the appropriate authorities can be notified and action taken. Where third parties have been authorised to gain access to the network, then such use should be inspected to ensure that proper procedures are being used.

6.13.3.2 Network pressure

If network pressure loggers or monitoring exists, then this can also be used to identify points at which pressure is regularly reduced for unexplained reasons. Further investigation could then take place to see whether unauthorised access is occurring.

6.14 BACKSIPHONAGE INTO THE NETWORK

6.14.1 Water quality risks

Backsiphonage is a reverse flow condition, created by a difference in water pressures, which causes water and associated contaminants to flow back into the distribution pipes of a drinking water supply from any source other than the intended one. Backsiphonage can occur from industrial, agricultural or domestic premises and can, in certain circumstances, lead to a significant contamination of the water supply network. If contaminated water or other liquids within the premises are in direct hydraulic continuity with the mains network, then under these reduced flow conditions, contamination can flow back into the network.

Backsiphonage can also occur within premises where it can lead to localised contamination of the private (non water supplier) plumbing network. This Section deals with mitigation of risks affecting the public supply network as a whole. Managing risks within premises generally uses the same approach, but is dealt with more fully in Chapter 7.

6.14.2 Risk mitigation

6.14.2.1 Risk assessment and prioritisation

Assessing, and seeking to mitigate, all potential backsiphonage risks within the network can be very difficult, time consuming and resource intensive. Realistically, in large networks it is probably impossible to eliminate all risks. However, using a risk assessment and prioritisation approach, the potential risks can usually be minimised to an acceptable level. This relies on a judgement being made about the frequency and magnitude of depressurisation events, the condition of the distribution network, and the type of premises within which the highest backsiphonage risks are likely to occur. For example, an area of the network in an industrial area and prone to regular mains bursts might have a higher priority for backsiphonage risk management than a network in good condition in an area primarily of domestic properties. However, backsiphonage contamination of the network can occur even in lower risk areas, so some level of mitigation work will be appropriate everywhere.

Risks can occur from a variety of types of premises. Risks can also be higher in premises where internal water systems are pumped to increase pressure, such as in large buildings, factories, hospitals or apartment blocks. Types of risk can include:

- Sewage and wastewater systems from both private and public buildings.
- Microbiological wastes from hospitals, clinics and other similar premises.
- Toxic or taste/odour causing chemicals from factories and industrial premises.
- Animal wastes and chemicals such as pesticides, from farms, abattoirs, stock yards and other agricultural sites.

6.14.2.2 Risk mitigation measures

The basic approach to reducing the risk of backsiphonage is proper design and maintenance of potable water plumbing systems to avoid the possibility of any cross linking with systems carrying waste water or other harmful liquids. Thus, even if reduced pressure occurs in the network, the likelihood of any contamination backsiphoning into the network is eliminated.

Design and maintenance of plumbing systems is a complex technical area which is often covered by national or regional legislation and/or official guidance. The responsibility for effective management of backsiphonage risks normally lies with the consumer or owner of the premises. However, water suppliers, in conjunction with government and other stakeholders, have a general responsibility to promote awareness of the importance of backsiphonage prevention and management. More information is given in Chapter 7.

The most cost effective approach for this relies on a combination of education and awareness, certification, inspection and legislation. The precise balance between each will be a matter for the appropriate authorities in conjunction with water suppliers and will vary depending on a range of local factors, including available resources, institutional arrangements and legal requirements. However, it can include:

- *Education and awareness*
  Education and awareness of the risks of backsiphonage are an important part of any risk management strategy. Many consumers and building owners are not aware of the risks, either to their own system or the local distribution network. Therefore providing practical information on the risks and ways that

they can be avoided is important. The best way to communicate with building owners will vary from country to country. However, developing partnerships with other relevant groups, such as regional and health authorities, national and local plumbing associations, plumbing retailers, builders, industrial trade associations, farmer groups and many others, is likely to be of significant value.

- *Certification*
  In some countries either voluntary or mandatory certification systems have been introduced which require that all plumbers have to be trained and certificated as competent to carry out various plumbing activities. This often also requires some form of continuous professional development to make sure that their knowledge is kept up to date.
- *Inspection*
  In some countries there are requirements, often backed up by legislation, which require that new plumbing installations, or those which are substantially altered, must be notified either to the water supplier or other authority. This allows for inspection to check for any backsiphonage or other plumbing risks. The resources available to check every such installation are usually limited, so in practice most water suppliers will use some form of risk based ranking to focus available inspection resources on those installations which are likely to pose the biggest risk. Additionally, some water suppliers have a programme of random inspection of high risk premises to ensure that plumbing systems do not pose an unacceptable backsiphonage risk.
- *Legislation*
  In a number of countries there are various legal requirements which help to minimise the risk of backsiphonage. Development of such legislation is a matter for Government, but can include:
  - Certification of plumbers.
  - Minimum standards of plumbing materials, design and installation.
  - Notification of new or altered plumbing systems.

6.14.3 Typical control points

Since implementation of mitigation measures for water quality risks associated with backsiphonage are outside the responsibility of water suppliers, determining the most suitable control points can be difficult. In practice, potential control points could include:

- Number of plumbers with appropriate training/certification.
- Number of high risk premises identified/inspected.
- Plumbing suppliers providing consumer/user information on backsiphonage risks.

6.15 NETWORK DESIGN AND WATER QUALITY

6.15.1 Introduction

In addition to the vital provision of water for human consumption, water distribution networks normally serve many other purposes, including provision of water for washing, sanitation, irrigation and fire fighting. Networks are usually designed and sized to meet peak consumer demands for all planned purposes, but this can be balanced to some extent by the use of treated water storage reservoirs. The design of distribution networks is primarily a hydraulic engineering issue, but there are important water quality issues which need to be taken into account as part of a WSP when designing a new network or improving/extending an existing network.

When designing a piped system, the aim is to ensure that there is sufficient flow and pressure at the point of supply to provide an acceptable service to the consumer. In many regions of the world this goal is not yet achieved, but, whatever the level of network development, water quality considerations should be included in the design of new or extended networks.

6.15.2 Water quality risks and their mitigation

There are a number of network design risk factors which can lead to a deterioration of microbiological or chemical quality. These have to be balanced against hydraulic, engineering and other local factors, but it is important that water quality is properly taken into account in the design and implementation process. Factors to consider include:

6.15.2.1 Excessive capacity

The system should not have excessive capacity which will result in long transit times, unless this excess capacity is required to meet a known increase in demand in the near future.

6.15.2.2 Low flow, dead ends and loops

Dead ends and loops in the system may cause problems by creating long residence times and areas where sediments can collect. Ideally, such situations should be avoided, but in practice this is not always possible. Similarly, sections which might experience low flow should be as short as possible. The possibility of tidal flows (change in flow direction) in loops should be avoided since it may disturb any deposits in the pipes.

6.15.2.3 Emergency cross connections

In many systems there are emergency cross connections which are normally closed by a valve, but which can be opened in an emergency to maintain flows within the network. However, this can also create areas where sediments can occur. Where such cross connections are present they should be noted on records and strict procedures be put in place to ensure that the valves are only opened when properly authorised. Additionally, the valves should be opened periodically, under carefully controlled, conditions to minimise the risk of sediment accumulation.

6.15.2.4 Pressure management

Pressure at any point in the system should ideally be maintained within a range whereby the maximum pressure avoids pipe bursts and the minimum ensures that water is supplied at adequate flow rates for all expected demands. This may require pressure boosting at strategic locations in the network. Situations that may give rise to transient low or negative pressures should always be avoided, since this could result in microbial or chemical contaminants being drawn into the pipe. Hydraulic models can be used to identify where, when and how negative pressures may occur. Such situations may occur, for example, where there are properties on high ground, remote properties at the end of long lengths of pipe, demands that are greater than the design demand, or pipes of inadequate capacity. Preventative measures, such as system reinforcement, may then be identified and implemented.

6.15.2.5 Access for maintenance

When designing a network, as far as possible it is important to incorporate fixtures that can be maintained with minimal disruption to normal flow regimes, using hygienic operating and maintenance practices. This might include, for example:

- Installing by-passes for devices such as flow meters and pressure-reducing valves that allow them to be taken out of service for maintenance.
- Designing valve chambers to be large enough to allow maintenance and replacement and are well drained to reduce the risk of contaminant ingress.
- Installing injection points at critical locations to facilitate emergency disinfection in the case of contamination.

6.15.2.6 Network zoning

It is increasingly common best practice to divide networks by valving into discrete management zones. For example, setting up so called *"district meter areas"* for leakage control purposes, where a group of, say, 1000–2000 properties is supplied through a single flow meter. However, as well as improved hydraulic management, zoning can lead to water quality benefits and/or disbenefits and these need to be considered and managed before implementation. Benefits may include easier containment of water quality contaminants after an incident, reducing the extent and complexity of mixing in distribution, or easier design and interpretation of water quality monitoring programmes.

Potential disbenefits may include increased flow velocity which might lead to resuspension of deposits, decreased flow velocity which might increase the rate of deposition, or potential increase/decrease in water age. It is, though, possible to limit the extent of these adverse effects by careful siting of the closed valves to minimise the creation of new dead ends with zero flow or long lengths with very low flow. Management of dead ends is particularly important when the boundary valves are opened. On these occasions, poor quality water may be carried into the neighbouring zone. It is therefore good practice to clean the dead end lengths on each side of the boundary valve before the valve is opened. Washouts installed close to the valve on each side will facilitate this process.

6.15.3 Typical control points

The most likely control points are regular checks on procedures to ensure that water quality considerations are taken into account as part of network design, and that design staff have appropriate training and awareness.

6.16 NETWORK MATERIALS AND WATER QUALITY

6.16.1 Water quality risks

Treated water conveyed through a piped network is exposed to numerous surfaces. It is important that no materials placed in contact with the drinking water in the network promote microbial growth or leach any contaminants into the water that can support microbial growth or are otherwise harmful. Thus, if not chosen with care, many different materials used in networks which are in direct contact with treated water can have a significant adverse influence on water quality.

In deciding which materials and fittings to use, a number of engineering, procurement and design issues have to be considered, but these must also be assessed against any potential microbiological or chemical risks. It is therefore important that water quality issues are taken into account as part of a WSP when designing and procuring materials for a new network or improving/extending an existing network.

The materials which need to be considered include pipes, as well as materials used to line or joint pipes. This also can apply to those parts of network fittings (e.g. valves) where material surfaces are exposed to the treated water. Additionally, materials used for the construction of treated water storage facilities, including jointing materials, can give rise to problems. The risks do not just apply to new or extended distribution networks, but also to network repairs or mains rehabilitation, especially when, for example, pipes are being relined with new material such as cement, plastic inserts or other material. Some of these problems do not necessarily become apparent immediately and some may result from pipe materials or linings installed many years ago.

Assessment of the water quality risks can be complex and require considerable local knowledge of the network and its hydraulic characteristics, but could include:

- Leaching of harmful organic or inorganic substances, or materials which can cause adverse taste/odours.
- Reaction between certain material (e.g. those containing phenolic material) and residual chlorine or other disinfectants, thus causing taste/odour problems.
- Leaching of alkaline materials from cement pipes or linings in low alkalinity waters, thus increasing pH.
- Leaching of material which promotes microbial growth.
- Corrosion, particularly of iron pipes and fittings, which can lead to discoloured water.

6.16.2 Risk mitigation measures and control points

In many countries use of pipes and material in contact with treated water is covered by special legislation which restricts the use of certain materials, or requires that only tested and approved materials are used in direct contact with water. It is thus important that design and procurement procedures for pipes and other network materials include measures to protect water quality, including any current legislative requirements. These procedures must cover not only pipes and equipment purchased directly by the water supplier, but also those used by engineering or other contractors.

Suitable control points are likely to be based on regular checks to ensure that these procedures are adequate, up to date, and being complied with. Where necessary, water suppliers may also wish to check that the quality control processes, used by manufacturers to make pipes or other network materials, ensure that procurement criteria are consistently met.

6.17 STRATEGIC TRUNK MAINS AND WATER QUALITY

6.17.1 Water quality risks

Trunk or strategic mains are large diameter pipes (usually $\geq$200 mm diameter) which are used to transport water from the water treatment works to a local distribution network. Normally there are very few, if any, consumers fed directly off these mains. Such mains are also used by bulk water producers to deliver water to local water suppliers which provide a direct service to consumers.

If designed and operated correctly, such mains are not a major source of water quality problems, but there are long term water quality risks which need to be taken into account. Large diameter mains are usually less easy to maintain than smaller diameter pipes. This is because they usually have few, if any, access points and the size of the main also often means that any work can be more complex and take longer to complete. Thus, large numbers of consumers can be affected if trunk mains have to be taken out of commission for repair or maintenance.

Although water quality risks tend to occur less frequently in trunk mains than in local networks, problems can still occur, particularly in older assets, and can have significant consequences because of the large numbers of consumers potentially affected. These include:

- Depending on the design and operation of the network and the quality of source water, sediments can build up in trunk mains over time, which can then lead to risks of water discolouration if disturbed.
- Although normally less vulnerable to risk of contaminant ingress, this still can occur through defective fittings or corroded pipes.

6.17.2 Risk mitigation measures

Because of the significant consequence of water quality problems with trunk mains, risk mitigation measures must be assessed, although the costs may mean that these need to be phased over a long time period as part of an asset management plan.

Ideally, if resources permit, trunk mains should be designed so that they can be taken out of use for repair or maintenance with the minimum possible impact on consumers. This could be secured either through duplication of the main or providing an alternative source of water into the local supply networks. If this is not possible then well rehearsed emergency procedures should be in place to make sure that any problems can be resolved in the most efficient way.

Regular inspection of the trunk mains should be carried out to detect any water quality risks or evidence of asset deterioration. These inspections should be recorded and any problems resolved as part on an asset management/water safety plan. Where necessary, plans should be made to clean the mains at appropriate intervals, taking into account the impact on consumers.

If the trunk main is owned and operated by a bulk water supplier, then there should be clear definitions of responsibility between the water provider and water distributor, covering areas such as water quality at point of handover, resolution of problems, asset management and emergency procedures.

6.17.3 Typical control points

The most likely control points are:

- Water quality monitoring at the end of the trunk main before it enters the local distribution network.
- Regular checks on asset integrity.

6.18 NETWORK MODELLING AND WATER QUALITY

6.18.1 Introduction

There is a range of network modelling software available with various levels of sophistication. Used appropriately, this can provide a valuable tool to support water safety plans and other aspects of water supply management. However, they require considerable skilled resources, reliable data, and well managed distribution networks to be fully effective. Network models are underpinned by hydraulic analysis and their primary use is for hydraulic management of the water supply network. However, they can also be used to support water quality management, provided that they are used with care and with due regard to any constraints or limitation on output data. Some more sophisticated models can include aspects of water quality modelling as well as quantity, but simple hydraulic models can still be of real value in managing water quality. In more sophisticated situations, models can be of even greater value when they are linked to GIS maps of the supply area, showing features such as asset location, land use, topography, soil type and links to customer databases.

6.18.2 Constraints and limitations to use

Network models require considerable resources to establish and maintain. They can provide very valuable information, but only if sufficient resources are available to ensure that the model design and operation, data inputs and management of the network itself remain effective. A network model is a substantial investment for any utility, so it has to be planned and managed with care. Factors to take into account include:

- Clear long term plans for model design and the way it will be developed, used and maintained in the future.
- Initial and ongoing input of reliable data on network assets and location. This should include robust procedures for updating the model when planned operational changes are made, for example new valves or network extension.
- Model validation to ensure that it reflects the real operation of the network. The involvement of all relevant staff, including particularly operational teams, can help to validate model outputs.
- Good and up to date input volumetric data covering water inputs from works, demand and diurnal variation.
- Strict procedures for network management. This particularly includes valve operation and other significant changes affecting system hydraulics, which should only be made as part of a permit system and with the knowledge of the network modelling team. The model will not provide valid outputs if those operating it do not know the up to date status of network assets at all times.

6.18.3 Water quality management uses

Reliable network models can be used for a variety of water quality management purposes, including water safety planning. The precise way and purpose in which it is best used can only be determined by the water supplier, but it could include:

6.18.3.1 Assessing the impact of new sources or water mixing

Bringing in new sources to existing networks or mixing different sources within the network can have many benefits, but also a number of significant adverse water quality impacts, if not properly managed. The use of network models is very important in effectively managing these risks.

6.18.3.2 Managing water quality deterioration

Water quality can deteriorate significantly within the network particularly related to factors such as biofilm growth, disinfection byproduct formation, tastes/odours and sediment deposition. Much of this relates to the age of the water. Use of network models can help to identify ways to minimise transport time to consumers, thus mitigating some of the worst impact.

6.18.3.3 Planning booster disinfection

In systems with long water distribution times it may be necessary to install booster disinfection within the network to maintain microbiological quality. Network models can help to plan the best location for this.

6.18.3.4 Maintaining network integrity and pressure

The risk of contamination of the distribution network through ingress increases significantly when pressure is low or zero. Equally, pressure surges can lead to disturbance of mains sediments or entrained air. Good network models can improve the way that hydraulic pressure is maintained in a balanced way across the network, thus minimising the risk of ingress of contaminants or other water quality problems.

6.18.3.5 Assessing consumers impacted by water quality problems

During water quality incidents, network models can support the protection of public health, for example by identifying those most at risk from contamination events and thus potentially requiring advice to boil water or take other measures. For events such as bursts or discolouration, models can identify those consumers most likely to be affected and thus assist in targeted consumer contact.

6.18.3.6 Analysing consumer complaint patterns

Using network models to analyse consumer complaint patterns can often help to pinpoint the root cause of the problem, thus leading to more effective targeted solutions (see Chapter 7). For example, the analysis of discolouration complaint data can help to identify those areas of the network that require priority in a mains cleaning or renewal programme.

REFERENCES

Ainsworth, R. (2004) WHO drinking water quality series. Safe piped water-managing microbial water quality in piped distribution systems. Published on behalf of WHO by IWA Publishing.

Bartram, J., Cotruvo, J., Exner, M., Fricker, C., and Glasmacher, A. (2003) Heterotrophic plate counts and drinking-water safety: The significance of HPCs for water quality and the human health. Published on behalf of WHO by IWA Publishing.

Boxall, J.B., and Saul, A. J. (2005) *Journal of Environmental Engineering* **131**(5), 716–724.

Davison, A., Howard, G., Stevens, M., Callan, P., Fewtrell, L., Deere, D., Bartram, J., (2005) Water Safety Plans Managing drinking-water quality from catchment to consumer, World, Health Organisation, Geneva, Switzerland.

Hayes, C.R., (2010) Best Practice Guide on the Control of Lead in Drinking Water, IWA Publishing, London.

Husband, S. and Boxall, J.B., (2010) *Journal of Environmental Engineering* **136** (1), 86–94.

Tokajian, S. and Hashwa, F. (2003) Water Science and Technology, **47** (3) 229–234.

Vreeburg, J. and Boxall, J.B. (2007) Discolouration in potable water distribution systems: a review, *Water Research* **41**, No. 3, pp. 519–529.

Water Research Foundation (2010) Report 4088 - Impact of chloride/sulphate mass ratio (CSMR) change on lead leaching in potable water.

World Health Organisation (2004) Guidelines for Drinking Water Quality, 3rd Edition, WHO, Geneva.

WHO/UNICEF (2010) Joint Monitoring Programme for Water Supply and Sanitation, 2010 update.

Chapter 7

Developing a consumer water safety plan

Michael Storey

7.1 INTRODUCTION

Consumer WSPs (often known by names such as Risk Management Plans and Water Quality Management Plans) have been used in practice in a number of countries for many years. In most cases these have taken the form of quality management systems, and have been used in a range of public, health care and commercial settings such as buildings, hospitals, hotels and ships, as well as for recreational pools and spas. Formal or structured WSPs, as such, have been much less commonly implemented for individual domestic properties, although many of the key requirements have been promoted and adopted through various consumer advice initiatives for managing domestic plumbing systems.

The general principles of WSPs that apply to all drinking water systems also apply when developing consumer WSPs. These consist of three key components (1) system assessment (2) management and communication and (3) feedback and improvement (WHO 2009). The main practical difference, as far as water utilities are concerned, lies in the responsibility for this. In the vast majority of cases a water utility's jurisdiction and liability for water quality ends at the point of connection, property boundary, check-valve or meter. After this point, the primary responsibility for water quality normally rests with the owners and/or operators of the building/facility. However, in some countries, and for some aspects, the utility's responsibility for water quality may extend within buildings or even to the consumer's tap.

Thus, while the utility has a key role in promoting and advising on best practice for water quality management within consumer's premises, both directly or through relevant (e.g. plumber) organisations, consumer WSPs should be developed by the owners and/or operators of the building or facility. At the same time, WSPs developed by water suppliers should complement and support consumer WSPs. For example, utilities could include in their WSP a framework that ensures that measures they are taking to protect water quality is aligned with the measures being promoted and adopted for consumer systems.

However, it must be appreciated that local legislation covering water in buildings can be highly variable between countries. For example, in many countries there are legal requirements covering such things as plumbing materials, plumbing installations, and measures to protect against *Legionella* in cooling towers.

Thus, while many of the principles described in this chapter, and in many cases endorsed by WHO, are regarded as good practice, it is essential that utilities take into account national/regional legislation and institutional arrangements before providing appropriate support to consumers in developing WSPs for their systems.

This Chapter therefore provides general practical guidance to water utilities on how to work with consumers and relevant stakeholders to promote the adoption of consumer WSPs in both residential and commercial settings. In particular, the Chapter covers how to:

- Identify relevant stakeholders (e.g. Government and regulatory agencies, health authorities, plumbers, developers and consumers) and their roles and responsibilities.
- Assess and mitigate common risks that are associated with the consumer's water system.
- Identify consumer groups most at risk, including sensitive (e.g. immunocompromised) populations.
- Provide effective consumer communication and liaison tools.
- Assess water acceptability to consumers through feedback and surveys.

7.2 STAKEHOLDERS, ROLES AND RESPONSIBILITIES

7.2.1 Introduction

Effective internal and external stakeholder engagement and management is imperative for the development and implementation of catchment to consumer WSPs (See Chapter 3.8). This requires close co-operation and partnership between decision makers and those with relevant expertise, and is critical for ensuring the confidence of consumers, utilities and regulators alike. However, stakeholder engagement is particularly important in relation to consumer WSPs. Water suppliers will need to work closely with the relevant organisations responsible for different aspects of consumer water systems. Key stakeholders for water suppliers should include, but not be limited to:

- Regulators – government, environmental and public health agencies.
- Developers.
- Contractors – architects, engineers and plumbers.
- Manufacturers and retailers of plumbing materials.
- Consumers – building owners, employees, operators, residents and public.

Integral to good stakeholder management is good communication, as well as clear definition of roles and responsibilities. Individuals need to understand their role in water quality management and public health protection, and how this can be achieved through the implementation of a WSP.

One way of securing this could be through a series of consultations with key stakeholders, covering every stage of the design, construction, operation and maintenance process of buildings. This could include workshops to identify ways to facilitate effective development and implementation of a consumer WSP. It could also include collating best available information on current and historic plumbing practices in order to characterise the typical plumbing materials (e.g. whether lead pipes had been used), and other relevant aspects of water system design, in different parts of the supply area. As part of this consultation process it is important to understand the risk culture of the different stakeholders (See Chapter 2). The extent to which the organisations involved have already developed such a risk based culture will influence the success with which they can successfully implement a WSP approach.

7.2.2 Regulators

Depending on the country, there are a number of activities and requirements impacting water supply systems in buildings that can be subject to regulation. However, the responsibilities and institutional arrangements for setting, managing and enforcing such regulation will vary considerably. For example, regulatory frameworks could include building and plumbing codes, occupational and health requirements, and codes applying to operation of water systems (WHO 2010). These regulations can be administered by a range of governmental and non-governmental agencies and can be formalised through specific regulations, operating licence requirements or memoranda of understanding (MoU).

Local and regional government agencies can play a key stakeholder role, particularly in the situation where they own, or have direct responsibility for, the water utility. Government departments may also perform a regulatory function and/or provide expert advice to ensure the safe supply and use of water in buildings, for example by developing legislation relating to plumbing, or plumbing materials.

Public and environmental health regulators, as well as advising on or setting drinking water quality standards, may also be responsible for surveillance and auditing of water systems, and ensuring compliance with guidelines or codes of practice. In the event that these are breached, or the unlikely occurrence of outbreaks of disease, they may also play a key role in overseeing health surveillance and incident management.

7.2.3 Developers

Developers of new buildings or other facilities have oversight of the entire design and construction process, which must include an understanding of the operational requirements and end uses of water. It is their role to engage architects, builders, engineers and plumbers and ensure that appropriate design requirements and standards are adhered to.

7.2.4 Contractors

For the purpose of this Chapter, a contractor is a generic term used to describe any individual whose role is to design, construct or operate the water system within a building or facility. It is also their responsibility to be aware of relevant regulations and codes of best practice.

Architects work in partnership with engineers and other professionals and are responsible for building design. As such, they need to be aware of the requirements of the water supply, as well as end users. Engineers are responsible for translating architectural plans into building designs, taking into account structural integrity and ensuring compliance with building and plumbing standards. Engineers are typically responsible for final certification of building at project completion (WHO 2010).

Plumbers play an integral role in risk assessment and management of consumer plumbing installations. Good communication with them and their industry trade associations or bodies is imperative for best practice. A fundamental requirement is the use of quality assurance systems to ensure that plumbers are well trained and that they adhere to a code of good practice (WHO 2006). Contractors generally belong to professional bodies, associations and institutes, which should be engaged at all opportunities to disseminate information relating to WSPs. These functions can include:

- Development and advocacy of codes, policies and practices.
- Practical guidance on development and implementation of WSPs.
- Training and accreditation of members.
- Providing information on management practices.

7.2.5 Plumbing manufacturers and retailers

Plumbing manufacturers, importers and retailers have a key responsibility to ensure that they understand and fully comply with any legislative or best practice requirements in relation to design and materials used. They may also be able to include best practice advice for fitting and use with the product packaging.

In many countries property owners can buy plumbing fittings from a variety of retail outlets and carry out alterations to plumbing systems themselves. In such situations the involvement of retailers in selling only approved products and promoting best plumbing practice and water safety precautions can be very important.

7.2.6 Consumers

A consumer is any individual that uses, or comes into contact with, drinking water either directly or indirectly, and can include employees, operators, residents or any other water user on a particular site or building. These individuals are at the front line of water safety and may be the first to detect problems with the water supply system. For this reason, reporting by consumers of any observed changes in water quality or faults with water systems should be encouraged and mechanisms established to support such reporting (WHO 2010).

Consumer communication and engagement is a key part of any water utility function. This includes the information on the development and implementation of a WSP, including the responsibilities for different parts of the water supply system. It is also important to recognise the particular needs of sensitive residential and commercial consumers, including immunocompromised individuals and persons requiring specialised medical treatment. These are covered later in more detail in Section 7.4.

7.3 SYSTEM ASSESSMENT AND RISK MITIGATION

7.3.1 Introduction

There is a wide range of hazards and hazardous events that can impact on water quality in a consumer's water system. These can be of either chemical or microbiological origin and can impact both water safety and acceptability. They generally fall within five broad categories:

- Poor hydraulic demand and on-site storage.
- Plumbing materials.
- Water temperature.
- Cross-connection, back-siphonage, ingress.
- Plumbing repairs.

The principles of system assessment and risk management have been covered extensively in Chapter 3 and other key references. Many of the risks that are relevant to consumer systems are similar to those described in Chapter 6 for water distribution networks.

Actual plumbing practices and materials, as well as legislative requirements and institutional arrangements for managing buildings, will vary considerably between countries. This Section briefly summarises the main hazards and hazardous events that can occur, along with typical mitigation measures. Some examples of common water quality hazards and related causes, mitigation measures and responsibilities are given in Table 7.1. Much more detailed information is available in key references (WHO 2006, 2010). While the information provided should be regarded as general good practice, utilities will need to check carefully with appropriate stakeholders, the specific legislative and other requirements in their area.

Table 7.1 Common water quality hazards in consumer systems – summary examples of risk mitigation, roles and responsibilities

Hazard	Cause	Risk mitigation	Consumer Role	Utility Role
Heavy metals	Material leachate (e.g. lead, copper, cadmium, nickel).	Replace materials, reduce retention time and flushing.	Use only qualified and certified plumbers. Replace unsatisfactory plumbing where practical.	Minimise corrosion potential of distributed water as far as practicable. Provide communication material and advice, promote best professional practice.
Opportunistic pathogens	Microbial regrowth - high temperatures, nutrients, low hydraulic demand, disinfectant residual (e.g. *Legionella*).	Disinfection, reduce retention time, flush system and mechanical cleaning, limit nutrients and temperatures, where possible.	Develop water safety plan, microbiological testing, and maintenance programme.	Provide communication material and advice, promote best professional practice.
Chemical or microbiological contaminants	Vermin, ingress, cross-connection, back-siphonage (e.g. metals, organic and inorganic chemicals, pathogens).	Cover tanks, vermin protection, valve maintenance, system checks, maintaining appropriate pressure differential, chlorination, plumber accreditation and training.	Develop water safety plan, chemical and microbiological testing, maintenance programme.	Provide communication material and advice, promote best professional practice.

7.3.2 Poor hydraulic demand and on-site storage

Poor hydraulic demand and long retention times are often implicated in chemical and microbiological contamination of consumer drinking water and on-site storage systems. On-site storage systems can represent the final barrier prior to consumption, so effective risk management is vital. Storage for potable use can take a number of forms, and water can also be stored for non-potable purposes such as pools and hot tubs, cooling towers and evaporative condensers.

Alongside high temperatures and nutrient availability, poor hydraulic demand is one of the leading factors contributing to microbial regrowth. This can result in the proliferation of pathogenic and other bacteria as well as microbially induced dirty water events (manganese, copper, iron oxidation). Poor hydraulic demand can also concentrate chemicals that may have leached from plumbing materials. Particular attention should therefore be paid to dead-ends or dead-legs in a consumer's distribution system, where stagnation of water and concentration of contaminants may occur.

Poorly designed or maintained potable water storage tanks, for example if uncovered or damaged, can allow ingress of birds, small animals or insects. This in turn can lead to significant contamination of the water system.

Mitigation measures are often covered in plumbing regulations and codes of practice and, where appropriate, should also be included in any standard operating procedures for building management. Good design and operation of the system to minimise long retention periods are important. Where necessary, disinfection with chlorine or other disinfectants can help to prevent regrowth of organisms. Storage tanks or other sources of potential ingress should be inspected regularly to ensure that they are properly sealed and/or covered.

7.3.3 Plumbing materials

Inappropriate choice, or poor installation, of plumbing materials used in pipes, joints and other fixtures can lead to a range of potential contaminants entering the water system. Special attention should be paid to materials that come into direct contact with drinking water. International guidelines (WHO 2011) or local legislation (e.g. USEPA 1991) will set out maximum contaminant levels for chemicals which may be leached into water. In many parts of the world lead is one of the most common contamination risks, although copper, nickel and cadmium may also present concerns, particularly where corrosion is a problem. Organic substances can also leach from plastic pipes and fittings, flexible hoses, glues, adhesives or liners, and may directly impact on public health or aesthetic qualities of water. Indirectly, organic substances can also provide a carbon source that promotes the regrowth of microorganisms (WHO 2010).

Although treatment of the water distributed by the water utility can minimise the extent to which leaching of some metals can occur from plumbing (see Chapters 5.8.3 and 6.6), it is preferable to implement mitigation measures within the building or facility itself. These can take a variety of forms. Plumbing materials should conform to accepted quality and performance standards (WHO 2006) and should preferably be covered by local legislation. Risks from plumbing materials can be further prevented and minimised by education, training and certification of plumbers and others who install water systems. This can be reinforced by the adoption of quality assurance systems such as regulations and codes of best practice (WHO 2006). These may prohibit the use of certain hazardous materials and practices. Plumbing inspection services should be established, with responsibility to ensure that these codes are adhered to.

Wherever inappropriate plumbing materials have been used historically, for example as a result of old plumbing practice, improvement programmes should be established to reduce or stop the release of hazardous materials, for example by replacing the responsible components of the system. Where this involves large amounts of pipework and fittings this may need to be a staged process. In extreme cases, where there is a significant level of contamination at individual properties, the following should be considered (WHO 2010):

- Prevent the consumption of contaminated water (issue advice, label taps and fittings, supply bottled or tankered water).
- Flush the system and verify effectiveness by chemical analysis.

7.3.4 Water temperature

Increased temperatures in water systems can result in the regrowth of microorganisms, mostly heterotrophic bacteria. While the vast majority of these organisms are harmless, they can lead to unacceptable aesthetic impact (taste, colour and odour) in the water. However, a subset of these organisms, termed opportunistic pathogens, may present a source of public health concern. The most commonly known waterborne opportunistic pathogens are *Legionella* bacteria, which are responsible for respiratory infection, particularly Legionnaire's disease.

Outbreaks of Legionnaire's disease are most commonly associated with those hot water systems that are optimal for multiplication and release of the organisms through contaminated water aerosols of a certain size range. Particular attention should therefore be paid to cooling towers and recreational spas. However, other devices, such as fountains, respiratory therapy equipment, showers, misting devices and even car-washes have also been implicated in outbreaks of this illness and should also be carefully maintained by facility owners/managers.

Case Study – Department of Human Services, Victoria Australia (2001)

Risk-based legislation for the control of Legionnaire's Disease first came into force in Australia in 2001 as a direct result of an outbreak of Legionellosis that had occurred one year earlier at the Melbourne Aquarium in Victoria. This legislation mandated registration of all water-cooled air-conditioning systems through a comprehensive site-specific risk management plan (WSP). This, incidentally, was the first legally enforceable document of its type in Australia, and possibly internationally, based on a risk assessment and management framework.

A typical risk mitigation plan for *Legionella* by building owners or operators should therefore focus on minimising retention time of water in a system (e.g. avoiding stagnation in storage tanks and looping dead-end sections), pipe materials, mechanical cleaning and flushing, as well as disinfection. Additionally, water temperature should be maintained outside the range of 25°C–50°C, though temperatures higher than this may present a scalding risk and also need to be taken into account as part of the WSP. Temperatures should therefore be kept below 55°C where risk of scalding may exist, but not below 50°C (WHO 2011). More detailed information can be obtained from key references (WHO 2007).

Other waterborne opportunistic bacteria that are associated with illness amongst immunocompromised individuals, often in hospital settings (nosocomial), include members of the bacterial genus *Aeromonas*,

Mycobacteria and *Pseudomonas*, fungi *Aspergillus* and free-living protozoa *Acanthamoeba*. It is important to appreciate that these organisms are naturally occurring and, as far as practicable, efforts should be made to inhibit their growth in a distribution system (see Chapter 6.7).

7.3.5 Contamination through cross-connection and backsiphonage

One of the most significant hazards within consumer premises is when the design or operation of drinking water systems leads to ingress of non-potable water or other hazardous liquids. This can commonly occur through cross connections between potable systems and other systems, or backsiphonage. Backsiphonage can also present a major risk to the wider distribution network (see Chapter 6.13). While this can be caused by failure in plumbing systems, the most common cause is human error.

The risks of cross-connection are usually greater in commercial and industrial premises, which, in addition to potable water, often have more than one piped reticulation system that may transfer toxic substances or waste products. Other premises, including high rise buildings and hospitals, often use dual water systems for heating and other purposes, and these have also been implicated in cross-connection events. With the increased use of alternative water sources, including on-site grey-water and recycled water systems as well as rainwater tanks, cross-connection at the household level also presents an emerging risk. In practice, all alternative water supplies to consumer premises should be treated as potentially contaminated systems, and regular maintenance and inspection are essential.

It is therefore imperative that appropriate controls are put in place to safeguard both the consumer's, as well as the municipal, drinking water system. Typical best practice measures have been summarised in Chapter 6.13.2, but can include a combination of the following:

- Installation regulations and codes of practice that include systematic processes to reduce the probability of cross-connections.
- Materials codes and regulations that allow potable and non-potable water plumbing to be easily discriminated.
- Regulations that limit the legal installation and modification of plumbing systems to licensed individuals.
- Opportunities to apply pressure differentials in certain situations to ensure that, if cross-connections occur, they are from higher to lower quality water.
- Installation of backflow prevention systems to reduce the extent of hydraulic influence from any cross-connections that do occur.
- Operational checking and connection auditing.
- Education and communication to explain the need to avoid creating cross-connections.

7.3.6 Plumbing repairs and alterations

It is common experience among water suppliers that significant risks to water quality can occur when plumbing systems are repaired or altered by individuals who do not have the relevant experience or training to carry out such work. This can be a particular issue in private domestic dwellings. Mitigation measures mainly relate to ensuring that suitable advice is widely available to consumers. This should emphasise that, as far as possible, only qualified people should carry out plumbing repairs or alterations. Where simple repairs are to be carried out by the householder, this should be done with appropriate precautions to prevent water contamination. Such advice can be promoted by water suppliers, but also by many other stakeholders, including, for example, those who sell plumbing materials for domestic use.

7.4 SENSITIVE CONSUMERS

7.4.1 Introduction

Within any community supplied by piped drinking water networks, there will be particularly sensitive or vulnerable residential and commercial consumers. Those at greatest public health risk include immunocompromised individuals and persons requiring specialised medical treatment. Consumer WSPs should clearly identify these susceptible populations, particularly in child, health and aged care facilities. A detailed description of these groups can be found in WHO *Water Safety in Buildings* (2010). Other "*sensitive*" consumers include special commercial interests (e.g. food and beverage industries, nurseries, aquaria) who may be at risk of changes or fluxes in water quality.

The primary role of a water supplier is to provide, through the distribution network, safe drinking water that consistently meets statutory and other requirements for normal domestic purposes. While a water supplier cannot supply water that meets the specific needs of all particularly sensitive consumers, they should work closely with the relevant authorities and organisations that do have responsibilities regarding the safe management and use of water by such consumers.

7.4.2 Medical treatment

Within industrialised countries there is a sub-population of individuals that may have a less developed or compromised immune system and thus be more susceptible to infection from a number of sources. Included amongst the immunocompromised group are infants and the elderly, organ transplant recipients and others receiving immunosuppressive medication, patients undergoing chemotherapy, diabetics and HIV-positive persons. Collectively these individuals may be susceptible to infection by (opportunistic) microbial pathogens (see Section 7.3.3).

Due to the vulnerability of its patients, many healthcare facilities therefore provide additional water treatment, including filtration, disinfection, softeners and deionisers. Where patients are in intensive or critical care facilities, including cancer, transplant and renal wards, particular attention is also paid to drinking water, beverages and ice (WHO 2010).

The appropriate authorities should also give consideration to individuals receiving at home care. Renal dialysis patients effectively ingest very large volumes of water which means that they have much stricter water quality requirements than normal consumers. These persons are therefore more vulnerable to microorganisms, endotoxins and chemical contaminants, as well as chloramination of drinking water supplies. Other medical patients that require a higher level of water treatment include those on respiratory machines.

Water suppliers should work with local area health services to ensure that particularly susceptible individuals have been identified and that effective communication or contingency plans are in place when the quality or continuity of supply cannot be maintained. Where water is used for a medical or dental purpose, protection of the municipal water supply should also be guaranteed through the installation of appropriate backflow prevention devices.

7.4.3 Commercial interests

There is a range of commercial interests that may be susceptible to changes in water quality. These include:

- Food and beverage industry.
- Aquaria.

- Plant nurseries.
- Other industries e.g. hairdressers, recreational pools.

Food and beverage industries, such as premises where food or drink is processed, hotels and industrial premises, have the greatest potential to pose a serious risk to public health if contamination events occur (WHO 2006). Such facilities are often covered by specific public health legislation that is enforced by public or environmental health officials. As with medical facilities, water suppliers should work with the relevant authorities to ensure that appropriate advice is available, particularly during incidents affecting public water supplies.

7.5 CONSUMER COMMUNICATION AND LIAISON

As set out in Chapter 1.3, the Bonn Charter recognises that to meet the goal of "*Good safe drinking water which has the trust of consumers*", the water supplied has to meet objectives which go beyond simply compliance with vital health and statutory standards. It reminds us that there are three related fundamental objectives to which all those involved in the supply of drinking water should strive:

1. *"Access to good, safe, and reliable drinking water.*
2. *Water that is not just safe to drink but considered of good aesthetic quality by consumers; and*
3. *Water supplies in which consumers have confidence"*

Alongside good safe drinking water quality and continuity of supply, effective communication with consumers is one of the strongest drivers of consumer engagement and can lead to much greater consumer satisfaction and trust. The Bonn Charter also highlights the fact that one of the key requirements to ensure consumer trust in their water supply is openness and transparency, including, for example, that regulatory agencies should:

1. *"Ensure that the process of setting quality- and health-related standards is transparent and the choices made with regard to the level of risk considered appropriate are fully disclosed;*
2. *Establish verification systems to assess compliance with drinking water quality standards, drinking water safety plans, and to ensure that the results of water quality testing are valid; and*
3. *Ensure there is full reporting of compliance results and problems in a way that can be understood by all".*

Full and open communication to consumers on the quality of water supplied and compliance with regulatory and other requirements is thus a vital part of any water utility operation. The way that this can be most effectively carried out will depend on many local factors, but it should ensure as a minimum that all consumers can have ready access to up-to-date and easily understood information, through publications, notices, websites or other suitable means. Local sharing of best practice in this area by utilities and regulatory authorities can often be helpful (DWI 2009).

Such communication may go much further than simply reporting water quality data, for example by explaining how the quality of water is maintained by the utility through use of the WSP approach, describing plans to further improve water quality and highlighting how more information can be obtained if required (Figure 7.1). As set out in Chapter 3.6, it is even more important that effective communication takes place with consumers during water quality incidents.

Your water quality

We supply you with high quality, safe drinking water – treated to meet the *Australian Drinking Water Guidelines*.

We routinely monitor and test for up to 70 different characteristics, including taste, colour, odour, micro-organisms and chemical content.

From 1 January to 31 March 2011, Sydney Water tested 244 filtered water samples. One *Cryptosporidium* oocyst was detected in one 100 litre sample on 2 February 2011 from the outlet of the Orchard Hills Water Filtration Plant.

Sydney Water notified NSW Health immediately and re-sampled. No further *Cryptosporidium* oocysts have been detected.

For more detailed information, see the *Quarterly Drinking Water Quality Report* 1 January to 31 March 2011 or the summary at sydneywater.com.au

A full copy of the free report is available from our head office during business hours.

Why does my tap water sometimes appear cloudy?

The cloudy appearance can be caused by water of different temperatures entering the system. It is usually only air bubbles and is completely natural, particularly in the warmer months. NSW Health advises that the water is completely safe to drink.

When you fill up a glass, the water will clear from the bottom up and be completely clear in a few seconds. Sometimes it may take up to a few minutes.

Did you know?

You can listen to Waterwrap online.

Visit sydneywater.com.au/publications/newsletters.cfm

We speak your language

For a free phone interpreter service, call 13 14 50.

Contact us

Web sydneywater.com.au **Enquiries** 13 20 92 **Faults and leaks** 13 20 90
Postal address Sydney Water, PO Box 399, Parramatta NSW 2124

SW264 4/11 Printed on recycled paper

Figure 7.1 Communicating with consumers - Sydney Water's newsletter contains water quality information and is distributed quarterly with customer accounts

In many countries service level agreements (SLA) exist between water utilities and consumers, and these clearly state regulatory requirements for water quality, which may include acceptability issues, consumer (customer) service levels, continuity of supply, as well as pressure. Information on, and compliance with, such requirements should also form an essential part of any consumer communication strategy.

Consumer trust in their water supply can be enhanced further if water suppliers are seen to be actively engaging and working with other key stakeholders, such as government, regulatory and health authorities. As described in Chapter 3.6, such relationships can be particularly important when responding to water quality incidents (See case study below).

Case Study – *Cryptosporidium* and Sydney Water (1998)

Sydney Water's Annual Residential Survey showed that the trust residential customers placed in Sydney Water was impacted following the events of 1998 when *Cryptosporidium* was detected in the finished water supply and boil-water alerts were issued. It took three years for trust levels to rebuild after the incident. Openness and honesty in handling mistakes have an impact on trust. From 2003 trust measures again softened as the impact of drought made people feel concerned about the future of the city's water supply.

It is also important that water utility consumer communication strategies clearly set out the roles and responsibilities of consumers, and others, in relation to maintaining water quality within consumer's premises. The Bonn Charter suggests that the role of consumers is to:

1. *"Operate and maintain the water systems in houses in a manner consistent with the maintenance of good water quality. Among other things, this requires the use of appropriate materials; and*
2. *Be good citizens by ensuring that their behaviour minimises the potential for contamination of water sources to occur or for supply quality or reliability to be diminished."*

Thus, information provided by the water utility should include such things as relevant roles and responsibilities, simple advice on best practice for managing water systems in buildings, and where to get more help and advice.

7.6 CONSUMER FEEDBACK AND SURVEYS

7.6.1 Introduction

Consumer trust depends not only on the consistent delivery of good safe drinking water, but also how consumers perceive water quality. Consumer feedback and surveys therefore provide powerful tools to assess and ensure the acceptability of drinking water supplies. It is also an important aid to the development and implementation of a WSP. Consumer acceptability data can be collected from customers both passively, through consumer complaints, and actively, through consumer surveys.

7.6.2 Dealing with consumer complaints and inquiries

In addition to meter reading and billing issues, water quality, particularly acceptability issues such as discolouration, taste, and odour, can be a common cause of customer complaint or inquiry. Other causes are generally related to factors such as pressure and interruption to supply. All consumer contact requires timely and effective response by the water utility. Careful analysis of complaint data should also help to improve operational and maintenance practices as well as organisational strategy and planning including for WSPs (see earlier Chapters).

Water utilities should establish systems appropriate to the local circumstances, and have in place adequate resources, to receive and process consumer complaints. The use of consumer contact centres, either in person, although most commonly by telephone or internet, is often the consumers' preferred communication channel. In many utilities a single, well-publicised telephone contact number has been established to facilitate all consumer contacts. In recent years water utilities have also explored the use of social media (e.g. Twitter® and Facebook®) to monitor and respond to customer sentiment and complaint.

Most utilities will make use of "*decision-trees*" to guide their staff through a standard set of questions and responses, including those for water quality complaints, for example:

- Can you describe the problem?
- Is the problem at the cold tap?
- Are neighbours experiencing the same problem?
- Is there anyone working in the street?
- Has this occurred before? If so when?

In most cases, a well-prepared script and properly trained staff will be able to deal effectively with the vast majority of consumer calls. However, it is particularly important to have procedures that can detect and rapidly respond to a potentially more serious contact, for example suggestions that the water may have led to sickness, or a large number of related calls from the same area. Anything that appears more serious or unusual will need to be investigated further as a matter of urgency by technical staff since it might be the first indication of a significant water quality or other incident (see also Chapter 3.6).

For use in planning and WSP purposes, consumer complaint data also needs to be logged and categorised against a consistent set of definitions. For example, taste and odour complaints may be categorised as chlorine, earthy/musty, petrol/diesel, or other taste or odour. In some countries or regions there may be specific guidelines or standards available, or a regulatory authority may define the categories under which complaints should be reported. Additional categorisation, e.g. identification based on duration or timing of the complaint (seasonal, continuous, or repeating), will also significantly aid analysis (UKWIR 2006).

Water suppliers can analyse customer complaint data in several ways to help them identify problems and their causes. In the context of a WSP, the main objective of any analysis is to identify where there are spatial or temporal patterns that may point to particular hazards and the need for intervention at a treatment plant or part of the distribution system.

Analysis will either be via spatial or temporal techniques. Spatial analysis will map complaint data onto network plans to show whether there is any clustering effect of complaints. This can highlight where there is a particular problem area of a supply network, or where there are issues with the source and treatment of water to an entire system.

The objective of temporal analysis will be to determine if there are any changes of certain complaints over time and whether this correlates with other parameters. The following examples of analysis may be used:

- What proportion of complaints is repeated and successive?
- Is there a relationship between water quality samples and complaints?
- Is there any seasonality of complaints?

- Are complaints related to specific causes – for example by analysis of the time difference between event (e.g. bursts) and complaints?
- Number of complaints received in the same area in a single day?

The advantage of using consumer complaint information is that it is possible to accurately record all complaints to a water supplier. However, not all consumers will complain and there may be a relatively poor link between this and overall consumer sentiment. Typically, for every one complaint there could be up to 10 people who have experienced a similar problem but did not complain. Also, the end user may not always detect taste and odour issues, since some people have an impaired sense of taste and smell, and others do not regularly drink water from the tap or filter their water. Customer surveys in some countries have indicated that less than 60% of customers claim that most of the water they drink comes from the tap.

The recording of consumer complaints as the only measure of dissatisfaction may not adequately represent problems that are continuous and about which the consumer does not expect anything can or will be done. Also, relying solely on consumer complaints as a measure of the acceptability of the water supply may misrepresent the scale of problem or views of certain consumer groups. Consumer acceptability surveys, on the other hand, provide a means by which the views of a wide range of consumers may be collected.

7.6.3 Consumer Acceptability Surveys

The acceptability of water supplies to consumers can differ considerably both within and between utilities, and will depend on a wide range of cultural, historical, social, demographic and economic factors. Thus the definition of what is "*acceptable*" water quality is subjective and can only be identified locally and through consumer feedback.

The advantage of consumer acceptability surveys is that they can explore factors that have a significant impact on public perception of water quality, beyond those identified through complaints. Water suppliers can use this information to assess whether consumers object to a particular source of water or treatment process and whether demographic factors play a role.

The assessment of acceptability of a water supply by a consumer goes beyond the simple question "*Do you drink the water supplied to you*?" It is, therefore, important that the water supplier clearly understands what their consumers mean by the term "*acceptable water supply*" or discusses the issue of acceptability with the relevant stakeholders, including other water industry experts, regulators, and consumer representatives, before surveys are carried out.

In some countries it is common practice for water suppliers to regularly (e.g. monthly) contact a small number of their consumers to assess the acceptability of their water supply. This frequent contact allows the water supplier to respond quickly to issues which arise in the network, and gives rapid qualitative and quantitative assessment of the acceptability of their water, compared with a larger more detailed survey carried out less often.

Figures 7.2 and 7.3 provide examples of a continuous on-line consumer sentiment (tracking) monitor. In this case a continuous online tracking study of adults is used, with quotas placed on age and gender to ensure a sample that is representative of the general population. Study findings are used to evaluate specific water-related initiatives and communications.

Designing the surveys, for example deciding who to ask, what questions to ask, and how to summarise the results, requires particular experience that water suppliers may not have in-house. Water suppliers therefore often engage specialist companies to carry out larger consumer surveys on their behalf.

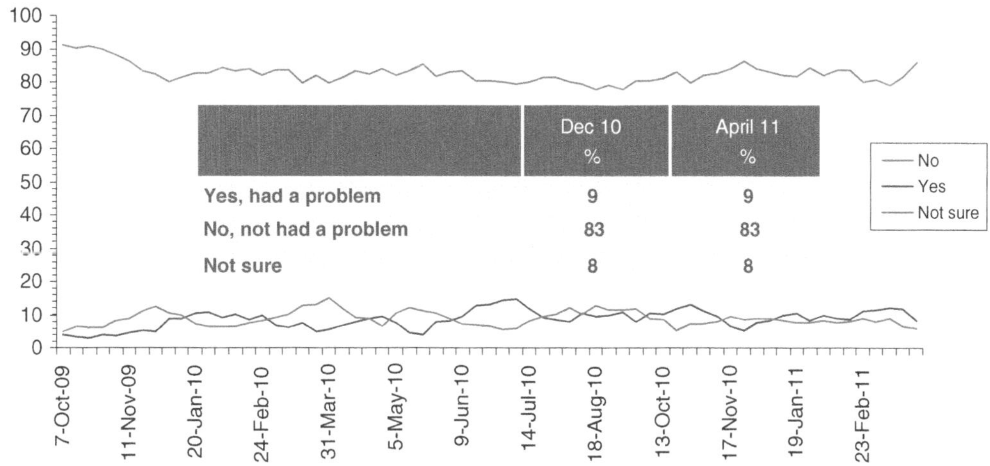

Figure 7.2 Online tracking survey. Question: Have you had any problems with tap water in your home in the past 12 months? (Base: All people - roll 4 weekly)

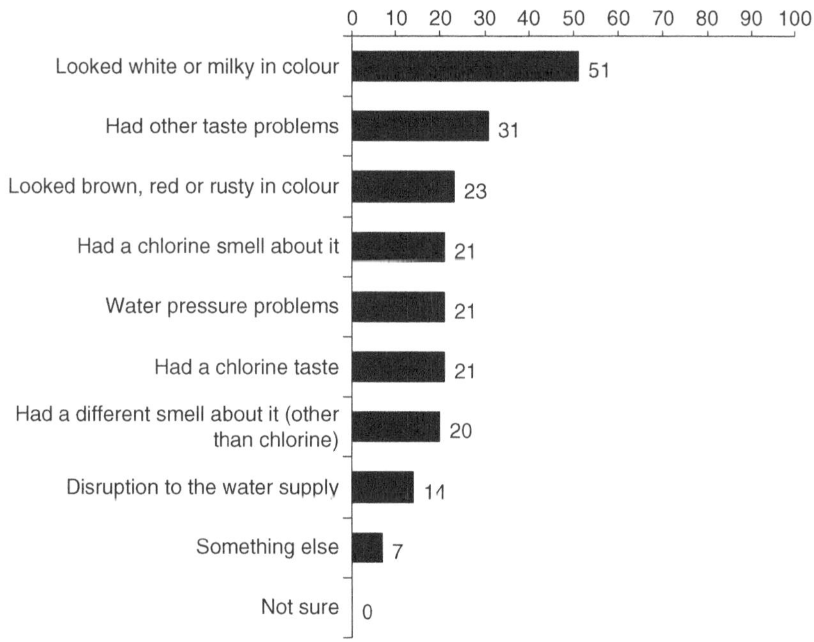

Figure 7.3 Online tracking survey. Question: What problems did you encounter with your tap water? (Base: People who have problems with tap water n = 42)

The results of these surveys should be reported to senior management and other executive staff, and can be used by water suppliers as a key indicator by which they can understand and develop relationships with their consumers. In some countries operating licence requirements also mean that these results are reported to regulators and the public.

A small number of studies have used observed or actual behaviour to indicate consumers' acceptability of drinking water supplies. Some consumers, for example, may not complain but will simply avoid using tap water and will turn to another source (i.e. bottled water, point-of-use treatment systems and boiling water). So-called "*averting behaviour*" has been one of the most popular approaches to evaluating drinking water safety in countries where it is a particularly significant issue.

Other research studies include sensory research and analysis that has its roots in psychophysics and the study of human sensory responses. They can involve the use of panels of people who are trained to be consistent in their evaluations, and thus act as analytical instruments under strictly controlled conditions. Sensory panel evaluation can be valuable when water sources or treatment are being modified and a minimal or undetectable impact on the sensory properties of the water is desirable.

Broader consumer research is also undertaken by water utilities to understand consumer views and perceptions of a water utility or some specific aspects of its products or services. This information is then used to inform the long term strategic direction of the organisation. These research studies are often carried out in several stages with participants investigating and exploring all aspects of the water industry to develop informed views of the issues. Again this type of research is carried out by specialist companies and requires the co-operation and sustained input from the participants and can be relatively expensive to carry out.

REFERENCES

DWI (2009) Report D5173/R1 - Good practice for communicating about drinking water quality. UK Drinking Water Inspectorate, London.

UKWIR (2006) Acceptability of water to customers: Report 07/CU/02/3, United Kingdom Water Industry Research, London.

WHO (2006) Health Aspects of Plumbing. World Health Organisation, Geneva, Switzerland.

WHO (2007) Legionella and the prevention of Legionellosis. J. Bartram, Y. Chartier, J. V. Lee, K. Pond and S. Surman-Lee (Eds.). World Health Organisation, Geneva, Switzerland.

WHO (2009) Water Safety Plan Manual. Step-by-step risk management for drinking water suppliers. World Health Organisation, Geneva, Switzerland.

WHO (2010) Water Safety in Buildings. World Health Organisation, Geneva, Switzerland.

WHO (2011) Guidelines for Drinking Water Quality, 4th Edition, World Health Organisation, Geneva.

Chapter 8
Global experience of WSP implementation

Tom Williams

8.1 INTRODUCTION

The inclusion of the Water Safety Plan (WSP) approach in the 3rd Edition of the Guidelines for Drinking Water Quality and the Bonn Charter for Safe Drinking Water, both published in 2004, catalysed a global shift in how water suppliers are managed. Their endorsement by governments and regulators, and implementation by water suppliers, have led to greater cooperation amongst all stakeholders, improved integration of practices and a more consistent and reliable approach to managing drinking water quality. Although management systems similar to WSPs, such as HACCP style approaches, were being implemented before 2004, their inclusion in internationally recognised and respected points of reference galvanised stakeholders involved in drinking water quality management to take concerted action.

International experiences have given rise to a better understanding of some of the critical success factors and bottlenecks to WSP implementation, mainly documented through case studies or anecdotal evidence. The benefits being realised are, in the main, consistent with the aims and objectives of WSPs, but patchy when considered in a global context. Cataloguing these experiences in a more methodical way to give a clearer understanding of the inputs, outcomes and impacts of WSP implementation is only a recent development. In years to come this will provide a much clearer picture on the extent of WSP implementation internationally, and their impacts.

This chapter will provide insight into some of the experiences of implementing WSPs over the past few years and aim to distil and synthesise some of the common challenges, critical success factors and opportunities for WSP implementation. In particular this chapter will address:

- Lessons learned from implementation: tangible benefits, common, recurring issues and challenges related to WSP implementation.
- Monitoring and evaluation: methodological approaches to assessing and improving WSP implementation.
- WSPs: integral to utility management: understanding the inter-linkages between WSPs and other utility planning and management processes.

- Impact at scale: understanding the enabling environment to support wide-spread, global implementation of WSPs.

The information in this Chapter is derived from publications, presentations and discussions at various global and regional conferences, as well as informal conversations with an international group of practitioners, policy makers and researchers. The contents of this Chapter should further support the practical understanding and implementation of WSPs according to guidance provided in this book.

8.2 LESSONS LEARNED FROM WSP IMPLEMENTATION

Chapter 1 of this book describes some of the potential benefits of WSP implementation (see Chapter 1.4.2). Some of these benefits address the explicit challenges that gave rise to the need to develop the WSP concept and some are more theoretical benefits. Since some benefits are not realised immediately, it has taken time to fully understand and interpret them. However, there is now a growing set of documented case studies and lessons learned, related to WSP implementation, that are demonstrating their impacts and benefits, from health (Dyke et al. 2007) to regulatory compliance (Gunnarsdottir & Gissurarson 2008), and from improved water quality (Mahmud et al. 2007) to economic benefits (Sancho et al. 2011). Furthermore, lessons learned from WSP implementation country and region wide are also providing valuable insight into some of the common challenges related to institutional, social and human resource capacity elements.

This evidence base acts as a further driver for the widespread implementation of WSPs and provides useful advocacy material for any practitioner championing the WSP approach within their utility or government agency. The need to implement WSPs in accordance with local circumstances means that the challenges and benefits vary. However, there are some commonly recurring issues that are worth detailing. This section will highlight a few areas of WSP implementation where key lessons have been learned.

8.2.1 Inertia: A common first hurdle

Chapter 2 goes into some detail regarding the institutional challenges, drivers and realities of WSP implementation in a utility. From top level management buy in to organisation wide understanding of the WSP approach, it is critical to nurture a common and consistent awareness of the WSP approach. Issues of perception and language are two of the big barriers to WSP implementation, particularly with regard to awareness raising and initiation. Understanding of the WSP concept by all levels of utility staff is critical, as is the terminology for commonly used words and phrases. It is, therefore, important that there is a common understanding of the basic principles and key terms. See Figure 8.1 for some of the key-words associated with WSPs.

Raising awareness of these key elements of a WSP will also prompt a recognition that the development and implementation of WSPs will include a number of existing practices (see Section 8.2.2 for more details on this issue). Another important message to communicate to staff involved in WSP implementation is that the WSP is not *additional* work, but is rather a new way of planning and undertaking their *current* work. There is a common perception that WSPs are developed alongside existing managerial and operational tasks and this can often lead to WSPs being viewed as a project, although, in the case of a pilot-scale WSP, one that is time limited. This view can lead to the WSP not being seen as a mainstream activity within a utility's operation and can create unrest or reticence amongst staff that see it only as additional work.

Figure 8.1 Typical words (*Word Cloud*) associated with WSPs

8.2.2 Where's the added-value?

Just as a WSP may appear to be extra work, it can also be perceived as an effort that has no intrinsic value. This may normally arise in situations where existing water quality is good and there are few water quality incidents. Furthermore, in some instances where WSPs have been introduced, it is found that many elements are already being implemented by a water supplier. The added value of a WSP for a water supplier is that it can more logically and comprehensively bring these elements together and help to identify where there may be gaps. This is certainly the case in Germany, where there was early scepticism regarding WSPs. This scepticism was due to an already high confidence in drinking water quality and management processes, which was underpinned by a well developed and highly regarded set of technical norms and standards being used by water suppliers and good levels of regulatory compliance. However, after a series of WSP pilot projects and consultations, it was found that the WSP elements of hazard analysis and risk assessment in particular were novel in relation to existing practices and deemed valuable contributions to managing drinking water quality (Schmoll et al. 2011).

8.2.3 How much will it cost me?

The costs of setting up a WSP are often cited as a barrier to implementation. However, these short term costs are thought to be much lower than perceived (European Union 2007). WSP implementation can save money in terms of reducing operating costs and better targeting investments for greater cost-benefit (Bartram et al. 2009). However, significant cost savings will only be realised in years rather than months. A number of utilities have reported cost savings on operational aspects related to WSP implementation. For example, in Uganda they reported a 30% reduction in their overall spending on water quality management as a consequence of implementing a WSP (Godfrey and Howard, 2004).

To date, only one study has evaluated the cost benefit of WSP implementation in a broader socio-economical context. Gerber (2010) estimated a cost:benefit ratio of 1:5.9 for WSP implementation in Palau (an island nation in the Pacific Ocean), which translates to an investment of US$ 0.2 million to implement and maintain a WSP and a benefits value of US$ 1.34 million over 20 years. A detailed global analysis of the cost-benefit of WSP implementation is a significant area which needs to be addressed in the future.

8.2.4 How to make friends and influence people

The importance of stakeholder cooperation has been emphasised throughout this book – from catchment authorities and farmers to health authorities and consumers. Good stakeholder cooperation comes from solid, sometimes formal, institutional agreements and honest working relationships between individuals. And this is where perhaps most early benefits of WSP implementation are being seen. Improved stakeholder communication and cooperation is something that may be difficult to quantify, but is undoubtedly critical in making catchment to consumer drinking water quality management effective.

The often cited WSP benefit of improved stakeholder cooperation can be difficult to translate into tangible benefits. However, Rinehold et al. (2011) described the impact of improved "*inter-agency*" communication as the more efficient use of time and resources, particularly with regard to addressing previously neglected but necessary actions. Conversely, insufficient buy in from a broad stakeholder group can foster a feeling of short-termism to implementation.

There is often frustration over the limitations brought about through lack of stakeholder engagement in fully implementing catchment to consumer WSPs. A common example of this occurs at the catchment level, where utilities, often with no or limited control over the catchments from which they extract water, feel powerless to make or influence improvements in the raw water quality. But the WSP approach should help to identify a range of potential solutions, ranging from short-term actions which help to ensure that treatment processes are more effective in treating raw water, through to longer-term efforts to influence catchment protection policy and regulation. The WSP approach can help to build up a robust evidence base for recommending and advocating change at the catchment level. Although certain elements of the water supply chain may not be in the control of utilities, it is important not to underestimate utilities' ability to influence the outcome.

8.2.5 Have we implemented yet?

"*When is a Water Safety Plan complete*?" is a frequently asked question. The obvious and most appropriate answer to this is "*never*". As previously stated, WSP implementation is an iterative process and should never be seen as a one-off project or time limited activity. But what this question often implies is: "*when can I say I have a fully implemented a WSP*?" In simple terms, a WSP can be said to have been implemented when all actions specified in a fully documented WSP are being undertaken and are accounted for, and it is subject to regular review and updating.

Making the leap from developing and documenting a WSP to implementation is often a challenge for some water utilities. Such challenges may be financial (e.g. no available finance to implement improvements), institutional (e.g. withdrawal of executive support for implementation) or simply due to a loss in momentum (e.g. other issues take priority), to name just three. So called "*Quick wins*" are, therefore, often important milestones in the early phases of implementation to ensure that efforts in planning are rewarded and discernable impacts may be seen.

It is critical to remember that a WSP should be implemented to suit the circumstances of the water supplier. A WSP can be as simple or complex, inexpensive or expensive, partially developed or comprehensively implemented, as it can or needs to be. However, core principles remain:

- The WSP should aim to bring about incremental improvements to aspects related to water quality; these may be small to start with.
- The WSP should identify short, achievable improvements, mid-term improvements which can be made with small additional resources and longer-term improvements which can be made with significant additional resources.

- WSP implementation is an iterative process and not an end in itself; it should represent and inform decision making processes related to drinking water quality.

When actions and responsibilities are undertaken as detailed in the WSP, complacency needs to be avoided and a culture of vigilance and continuous improvement should be encouraged. In avoiding complacency, regular monitoring and evaluation of WSPs is critical.

8.3 MONITORING AND EVALUATION OF WSPS

Monitoring and evaluation of WSPs, in an input and output context, are important to ensure that they are complete and effective, and also to highlight areas for improvement and progress. Mudaliar et al. (2011) suggests that, to date, most attempts to assess or evaluate WSP implementation are impaired, as evaluation procedures are often developed well after a WSP has been implemented. The consequences of this are a lack of baseline data and other collected evidence to enable comparison between pre and post WSP implementation. With this in mind, a number of organisations, including water utilities and research institutes, have begun to reflect on WSP implementation and develop assessment and evaluation frameworks based on lessons learned from implementation. In doing this, the challenge is to find a method that remains objective and is ultimately useful in providing a point of reference from which to make further improvements.

In this section, monitoring and evaluation of WSPs are reviewed within the context of output assessments (*outcomes and impacts* of WSP implementation), and input assessments (the *actions taken* to implement a WSP).

8.3.1 Output assessment

The main purpose of implementing a WSP is to consistently ensure drinking water quality in order to safeguard public health. However, at time of press there remains just one study (Dyke et al. 2007) that has demonstrated direct health improvements through WSP implementation. Health impacts are notoriously difficult to determine for water, sanitation and hygiene interventions, and, coupled with the time lag (which can be years) between implementation and sustained drinking water quality improvements, this lack of evidence is perhaps no surprise.

For this reason it is important, when considering the impact of WSPs, not to look immediately for the "*holy grail*" of health outcomes, but rather focus on intermediate benefits of WSP implementation. Such an outlook is critically important in establishing measurable milestones and progress towards WSP implementation. A framework for this approach has been developed by the United States Centres for Disease Control and Prevention (CDC 2011). This framework proposes a logic model approach that analyses the inputs, activities, outputs, outcomes and impacts associated with WSP implementation. This logic model is then applied to a WSP context within an assessment framework (see Figure 8.2) to determine institutional (e. g. improved stakeholder engagement), operational (e.g. improved efficiency of treatment processes), financial (e.g. reduced operating costs) and policy (e.g. recognition of WSP process in formal policy) impacts.

This framework can provide a good planning, implementation and assessment tool (see Table 8.1) for any supplier implementing a WSP, or even at a broader country level for establishing and assessing overall implementation targets for a country. Specifically the framework helps to indentify:

- The range and type of inputs required to implement a WSP.
- How process indicators (inputs, activities and outputs) have an effect on outcomes.
- The range and types of impacts of WSPs.

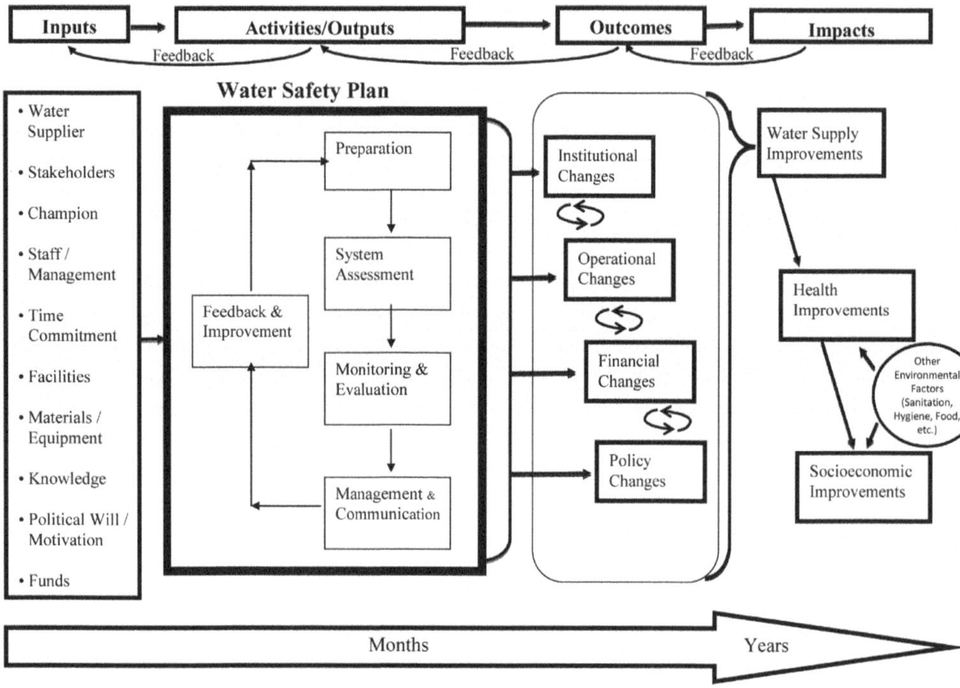

Figure 8.2 Conceptual framework to evaluate WSPs (CDC 2011)

Table 8.1 Application of the CDC conceptual framework to evaluate WSPs

How the framework can be used to support planning.	• Water utility wanting to develop a WSP can review previous WSP implementation experiences using the framework to identify the kind of inputs required to implement a WSP. • In a broader context, inform the development of national/regional capacity building programmes to support WSP implementation.
How the framework can be used to support implementation.	• The framework can be applied to determine milestones related to outputs and outcomes that will help to monitor the progress of implementation and also assess the relationship between inputs and outputs.
How the framework can be used to support assessment.	• A water utility can use the framework to assess a range of impacts that WSP implementation has.

The ultimate measure of effectiveness should, however, be set against local circumstances and reflect upon performance and water quality targets, for example health and statutory standards, consumer acceptability, quality reliability and consumer trust as described in Chapter 1.3. Just as a WSP should only be developed and implemented according to local circumstances, so should assessment of its effectiveness.

8.3.2 Input assessment

The aim of an input assessment is to report on the completeness and adequacy of actions required to implement a WSP. There have been a number of tools developed to assess inputs to a WSP (or similar drinking water quality management practices), mainly by individual utilities, research organisations and regulatory authorities. The usefulness of these tools is often restricted to the specific circumstances (e.g. specific national legislation or policy) in which WSPs are being implemented and reviewed. Some assessment tools are quite simple (e.g. in the form of a quick checklist) and establish some key actions that need to be undertaken to implement WSPs. Such tools are most useful where experience of WSP implementation is low and capacity to implement may be lacking. Whatever the complexity of the assessment tool, one overarching principle is constant; any input assessment should be used periodically, over time, to determine progress and drive improvements.

Australia was one of the first countries to produce a tool which aimed to assess and report on the strength of a water quality management system being used by a utility. This tool is called *Aquality* and was developed by the Water Services Association of Australia (WSAA) in 2006. The tool is specifically tailored to assess against a set of actions in the Australian Drinking Water Guidelines (ADWG) and can potentially be used for benchmarking purposes (Donlon et al. 2006). There are a total of 76 actions arising from implementation of the ADWG, for which Aquality has 187 measures (see Figure 8.3). During the first two years of application (2006-2008) the regular use of the tool led to a greater level of confidence and transparency in urban water quality and health (WSAA 2010).

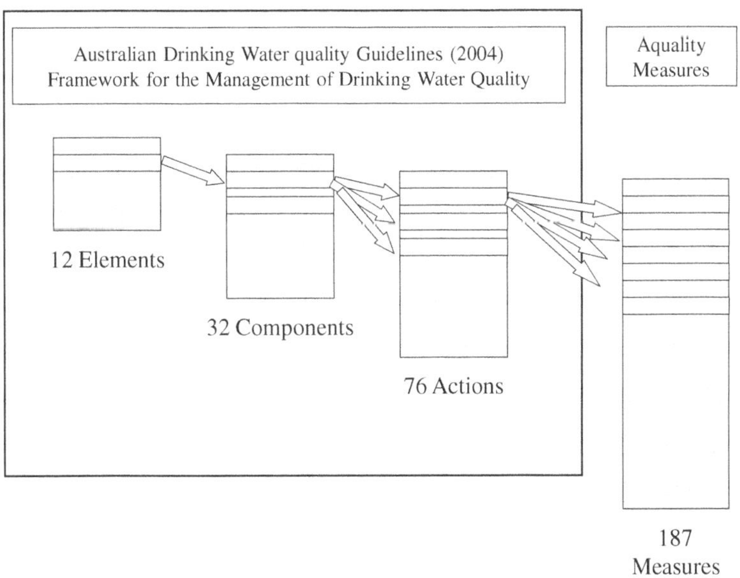

Figure 8.3 Relationship between Aquality and Australian Drinking Water Quality Guidelines (adapted from Donlon et al. 2006)

Critical to the assessment of WSPs is that the process is objective and drives improvement. National programmes, such as the Blue Drop System (BDS) used in South Africa (South African Department of Water Affairs 2011) as a regulatory tool to monitor the status of drinking water quality management, have been used to good effect in understanding the nationwide status and bottlenecks impeding WSP implementation (de Souza et al. 2011). The main criteria for the BDS are shown in Table 8.2 and are based on a system of weighted scoring against nine criteria. It is important to recognise that, since the programme commenced in 2009, the weighting score allocated to WSPs and asset management has increased from 0% to 15% in 2011, and at the same time the weighted score allocated to drinking water quality (DWQ) monitoring programmes, DWQ compliance and management of DWQ failures has reduced. This is to reflect the incorporation of some of these elements within the WSP approach and also to acknowledge a lesser reliance on DWQ compliance and monitoring for ensuring drinking water safety.

Table 8.2 Criteria for South African Blue Drop status

Criterion	Allocated weight (%)		
	2009	2010	2011
Water safety plan and incident management	0	5	15
Process control, maintenance and management skill	10	10	10
Water quality monitoring programme	20	15	15
Water quality sample analysis credibility	5	5	5
Submission of water quality results	10	5	5
Water quality compliance	35	30	30
Management of water quality failures	20	15	-
Publication of water quality performance	0	5	5
Drinking water asset management	0	10	15

The increase over time in the allocated weighting for WSP criteria has also given water suppliers greater time to develop and implement their WSP. This progressive approach to regulation provides utilities with ample time to fully understand, develop, implement and refine WSPs with the full support of the regulator.

Australia and South Africa provide a relatively mature and organised setting for WSP implementation which supports effective monitoring and evaluation. However, many countries, particularly in low and middle income regions, lack the resources and tools to establish such programmes. To address this need, the International Water Association (IWA) and the World Health Organisation (WHO) produced the WSP Quality Assurance tool. The aim of the tool is to enable the objective assessment of a WSP and thus systemically highlight areas where progress is being made and opportunities for improvement. It also helps WSP teams to develop and implement WSPs. For those water suppliers that have already implemented a WSP, it can also help to challenge complacency by ensuring that the WSP remains up to date and continues to be effective. The development of the tool was reliant upon widespread piloting with a range of utilities of different sizes, operating in different geographical regions. This piloting ensured that the tool could be applied and the assessment outputs would be relevant in a variety of circumstances. The added value of this tool is that it has been developed on a generic basis, without bias, and can be adapted to suit local circumstances.

8.4 WATER SAFETY PLANS: INTEGRAL TO UTILITY STRATEGY AND PLANNING

The South African Blue Drop System demonstrates the emerging importance of WSPs as a significant component of a water utility's overall strategic planning and operations. The nature of WSPs, which cuts across many operational and managerial aspects such as asset management, regulatory compliance and consumer engagement, gives further importance to the need to raise awareness of WSPs across utilities' operations and encourage its use in decision making.

The relationship between WSPs, drinking water quality compliance and capital investment planning is a growing area of interest. This is being seen, for example, in the United Kingdom where the economic regulator for the water and sewerage industry in England and Wales (Ofwat) recognises that "*Drinking Water Safety Plans are important procedures to cover when identifying and justifying asset maintenance programmes*" (Ofwat 2008). There is a similar model being applied in Australia (see Figure 8.4), where WSPs inform a 5 year water plan, which includes capital investment requirements.

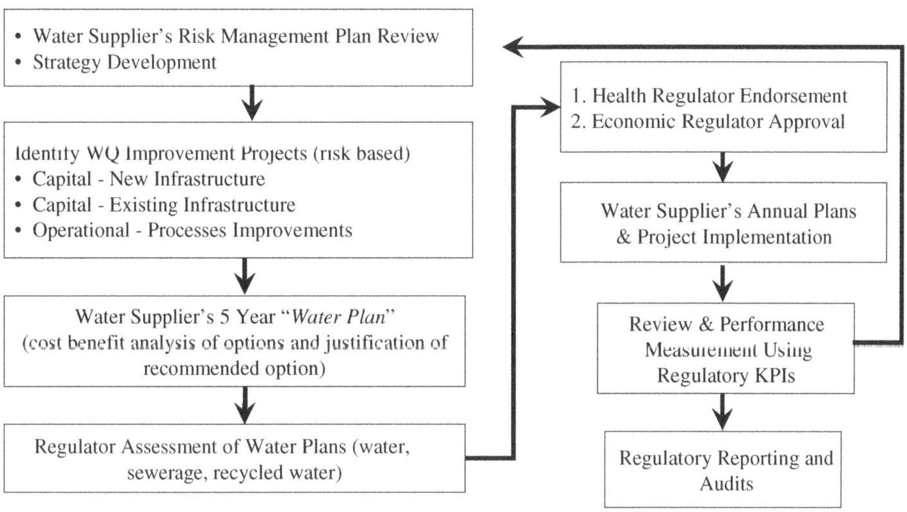

Source: Asoka Jayaratne, Yarra Valley Water

Figure 8.4 Role of WSPs in Australian regulatory structure

Such incentives for WSP implementation could readily be applied in a developing country context, either through a regulatory framework or through lending programmes, whereby implementation of a WSP, and the production of an upgrade/improvement plan within it, is recognised as a sound means to justify finance or credit. In many low and middle income countries, water utilities operate in perilous economic and institutionally unstable environments. Such circumstances often give rise to chronically inefficient and under-performing service providers, which ultimately lead to a water utility not being financially viable, giving rise to under investment and poor operation and maintenance. This is often called the "*Vicious Cycle*", whereby there is a recurring issue of under-investment, poor-performance and low revenue collection (Rinehold et al. 2011).

Turning this into a "*Virtuous Cycle*", whereby performance is improved and/or operating costs are reduced, is often advocated by non-revenue water (NRW) specialists. But the same could be applied for WSPs. In a number of circumstances, WSP implementation has led to a decrease in operating costs with the same or improved performance. Such cost savings could be used towards financing small improvements to operations, as determined by the improvement plan in WSP. But whilst this may help a utility to implement short-term, low-cost improvements, there still remains a bigger barrier for mid to long term improvements that normally require more substantial capital investments.

However, a WSP can *support* improved access to finance and credit to make such improvements in two ways (see Figure 8.5). Firstly, a water utility demonstrating sustained performance and service level improvements becomes a much more attractive proposition for donors and lenders, potentially opening up new avenues for finance and credit. Secondly, the exercise of producing an upgrade/improvement plan as part of the WSP process is rapidly becoming a robust and reliable mechanism to inform and justify capital investment requirements.

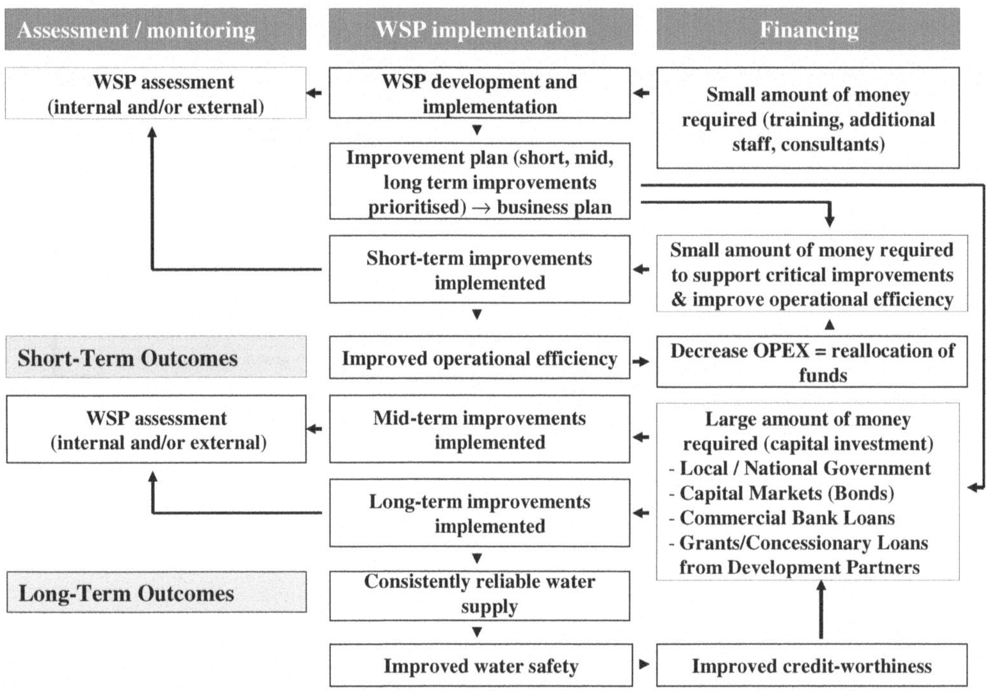

Figure 8.5 Conceptual framework for financing drinking water quality improvements using WSPs

8.5 IMPACT AT SCALE

The impact of WSP implementation on a global scale, as it relates to improving drinking water quality and safeguarding public health, can only be seen when large populations are served by drinking water supplied by water operators that are continuously assessing and improving their WSP. There are only a few countries that are at this stage, in a formal sense, through rigorous internal and/or external audits (e.g. United Kingdom, Australia, and New Zealand) but momentum is building. With emphasis on progressively incorporating WSP elements into national policy and best practice, incremental improvements to water quality can be realised. The spectrum of formality, regarding WSP adoption and implementation at a country level, is shown in Figure 8.6. Such a spectrum supports a concept of country readiness for WSP implementation by ensuring that national policy and/or regulation is commensurate with the status on the ground and is used as an incentive for WSP implementation.

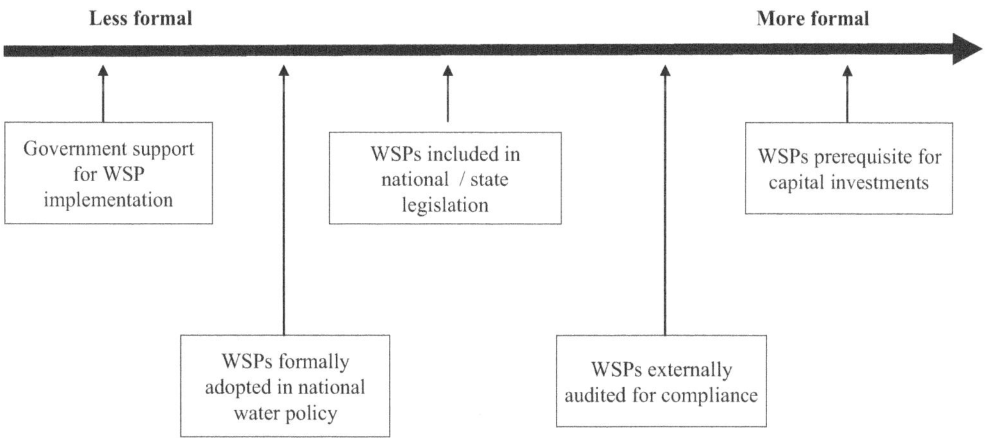

Figure 8.6 Spectrum of formality for inclusion of WSPs in national policy

At a country level, the pathway to widespread WSP implementation varies and is often dependent on key drivers. A recent publication from IWA and WHO, *Think Big, Start Small, Scale Up: A Road Map to Support Country-Level Implementation of Water Safety Plans* (WHO/IWA 2010) identified four main drivers for countries introducing WSPs:

- A decision by governments to encourage or require WSP implementation by water suppliers as a means to improve public health.
- Pro-active implementation of WSPs by water suppliers to improve performance, drinking-water quality, compliance or due diligence.
- Donor-driven requirements for project administrators to comply with international good practice.
- Promotion or support from professional and industry associations to encourage implementation at the individual water supplier and sector levels.

The rate of WSP implementation across a country or region can be dependent on what the main driver is. For example, in a country where a government encourages or requires WSP implementation, there is likely to be quicker undertaking of WSP implementation than in a country where WSP implementation is solely donor driven. This, of course, is dependent on a number of other variables, and in reality a combination of two or more of the above drivers is required for successful, widespread WSP implementation.

This is the case in Portugal, where WSPs have been implemented as far back as 2003 (Vieira 2007). Sector-wide, inclusive and pro-active WSP implementation has enabled Portugal to be a pioneer in the practice or WSP implementation. Vieira (2011) captured this maturation of WSP implementation within three progressive phases (see Figure 8.7), with the three "*pillars*" of progress being institutional settings, practical implementation and supporting mechanisms. What is clear from the Portuguese experience is that all stakeholders involved in the provision of drinking water quality need to play an active role in promoting, supporting and implementing WSPs. This is further supported in other countries, such as Germany, where a recent review of their national programme for WSP support confirmed: "*…the importance of, and need for, a proper dialogue and communication process with key stakeholders such as regulators, water suppliers, surveillance authorities and professional associations [for WSP implementation]*" (Schmoll et al. 2011).

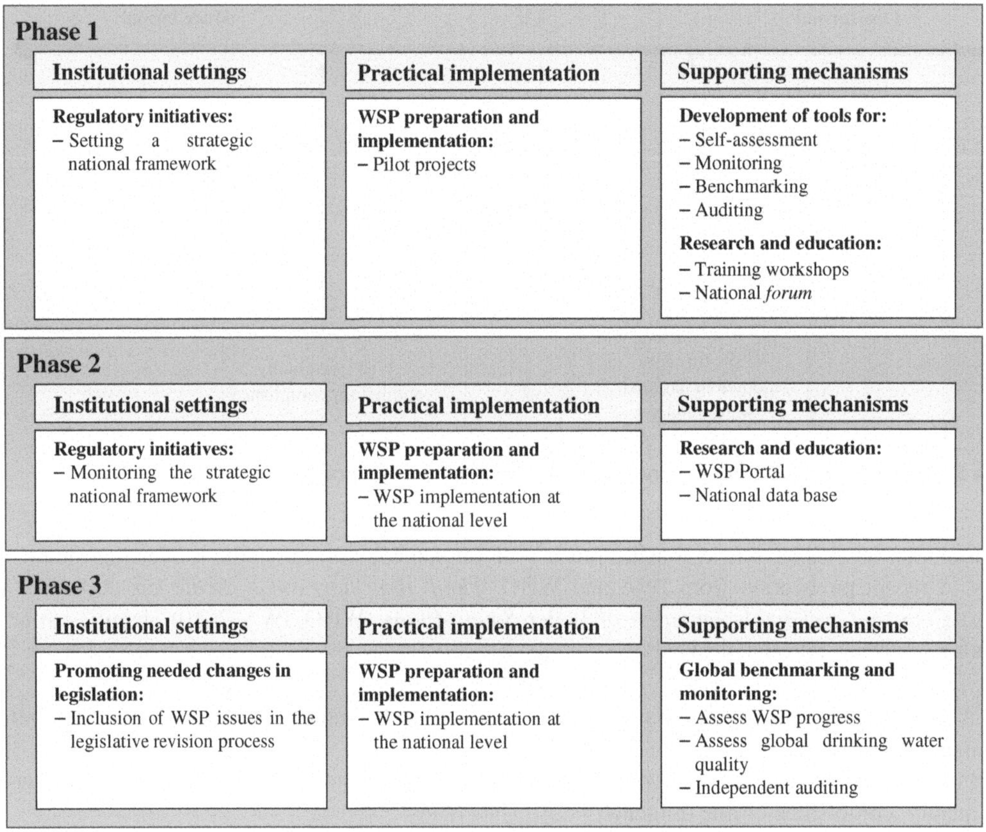

Figure 8.7 National strategy for WSP implementation in Portugal (Vieira 2011)

8.6 SUMMARY

Successful implementation of WSPs can improve drinking water quality, secure operational efficiencies and provide a robust framework to better target more sustainable capital investments. The realisation of health, operational and financial benefits obtained through WSP implementation has contributed to a growing evidence base that they are the most effective means to consistently providing safe drinking water. International experience to-date points towards some key lessons-learned:

- The practices included in the WSP approach are not entirely new, nor are WSPs "*additional*" work. They provide a new way of planning and an integrated, more consistent way to manage water supplies.
- Benefits are wide-ranging, not just health, but operational, financial and institutional. However, more analysis of cost-benefits and long term health impacts is required.
- Monitoring and assessment of WSPs needs a methodical approach and should drive continuous improvement.
- The role of WSPs as a regulatory tool can be effective provided a robust institutional setting is in place that supports and incentivises implementation by water utilities.

- WSPs are an integral part of a modern water utility's strategy and planning, providing a framework to inform a wide range of practices.
- WSPs have a key role to play in informing decisions regarding capital investments in the water sector, based on an outcome orientated approach to improving drinking water quality and safeguarding public health.

Given the above, the prospects for large scale WSP implementation making a significant contribution to improving the safety of global water supplies are good. Fundamental to this is the role of water professionals, not just as operators, technicians or engineers, but as custodians for safeguarding public health.

REFERENCES

Bartram J., Corrales L., Davison A., Deere D., Drury D., Gordon B., Howard G., Rinehold A., Stevens M. (2009) Water safety plan manual: step-by-step risk management for drinking-water suppliers. World Health Organization. Geneva, 2009.

Centers for Disease Control and Prevention (CDC) (2011) A Conceptual Framework to Evaluate the Impacts of Water Safety Plans Centers for Disease Control and Prevention.

De Souza P.F., Burgess J.E., Swart M., Naidoo V and Blanckenberg A. (2011) Web Enablement of a Water Safety Plan via the Municipal-based Electronic Water Quality Management System (eWQMS). *Water, Science and Technology: Water Supply* (In press).

Donlon P., Davison A. and Deere D. (2006) Sleeping Well at Night: Implementing Continuous Improvement in Water Quality Management using Aquality. *WSAA Journal*, Issue 6 November, 2006, 1–13.

Dyck, A., Exner, M. and Kramer, A. (2007) Experimental based experiences with the introduction of a water safety plan for a multi-located university clinic and its efficacy according to WHO recommendations, *BMC Public Health*, **7** (34).

European Union (2007) Support for the development of a framework for the implementation of water safety plans in the European Union (*http://circa.europa.eu/Public/irc/env/drinking_water_rev/library*)

Gerber F (2010) An Economic Assessment of Drinking Water Safety Planning Koror-Airai, Palau. 2010 SOPAC Technical Report 440.

Godfrey S. & Howard G. (2004) Water Safety Plans: Planning urban piped water supplies in developing countries, WEDC.

Gunnarsdottir, M. J. and Gissurarson, L. R. (2008) HACCP and water safety plans in Icelandic water supply: Preliminary evaluation of experience, *Journal of Water and Health*, **6** (3), 377–382.

Khatri K., Iddings S., Overmars M., Hasan T. & Freshwater, A (2011) Lessons learned from implementing the Drinking Water Safety Planning Programme in the Pacific. *Water Science and Technology: Water Supply.* (In press).

Mahmud S.G., Sk. Abu Jafar Shamsuddin, M. Feroze Ahmed, Annette Davison, Dan Deere and Guy Howard (2007) Development and implementation of water safety plans for small water supplies in Bangladesh: benefits and lessons learned. *Journal of Water and Health*, **5** (4), 585–597.

Mudaliar, M.M, Walker, M., & Tully, D. (2011) Success or Failure: Demonstrating the effectiveness of a Water Safety Plan. *Water, Science and Technology: Water Supply* (In press).

Ofwat (2009) Setting price limits for 2010-15: Framework and approach. http://www.ofwat.gov.uk/pricereview/pap_pos_pr09method080327.pdf.

Rinehold, A., Corrales, L., & Gelting, R. (2011) Water Safety Plan Demonstration Projects in Latin America and the Caribbean: Lessons from the Field. *Water, Science and Technology: Water Supply* (In press).

Sancho, R.; Calvinho, A.; Fontinha, P.; Costa, L. (2011) Benefits, effectiveness and assessment of the Water Safety Plan – Case Study, Águas do Algarve, SA. *Water, Science and Technology: Water Supply* (In press).

South African Department of Water Affairs (2011) Blue drop requirements. Access via http://www.dwa.gov.za/dir_ws/DWQR/Subscr/ViewNewsDoc.asp?FileID=127.

Schmoll O, Castell-Exner C. & Chorus I (2011) From International Developments to Local Practice: Germany's Evaluation and Dialogue Process towards Water Safety Plan Implementation. *Water Science and Technology: Water Supply.* (In press).

Vieira, J. M. P. (2007) Water Safety Plans: Methodologies for risk assessment and risk management in drinking water systems. In Water in Celtic Countries: Quantity, Quality and Climate Variability (Ferreira, J. P. L. & Vieira, J. M. P. eds.). IAHS Publ. **310**, 57–67.

Vieira, J. M. P. (2011) A strategic approach for Water Safety Plans implementation in Portugal. *Journal of Water and Health*, **9** (1), 107–116.

Water Services Association of Australia (WSAA) (2010) Submission to National Water Commission 2011 Biennial Assessment November 2010 http://www.nwc.gov.au/resources/documents/WSAA.pdf.

WHO / IWA (2010) Think big, start small, scale up: A road map to support country-level implementation of water safety plans.

Appendix 1

The Bonn Charter for Safe Drinking Water

1. INTRODUCTION

1.1. Purpose

The reliable supply of good, safe drinking water is fundamental to a healthy community and to its economic development.

Its delivery requires a comprehensive understanding of contamination risks and effective control of those risks. It also requires that robust quality standards be in place and that systems be implemented to verify that safe water is produced. Systems established should be transparent. The provision of safe drinking water demands the participation of all stakeholders.

This Charter provides a framework for establishment of such systems, including assessment and mitigation of risks at all points in the supply system from catchment (Note 1) to consumer, and performance verification based on predetermined standards and controls. It also provides guidance as to the institutional roles that should be established and the benefits to be achieved through close cooperation.

The Charter is offered as a framework that could be applied universally by appropriate organisations as a basis for reliably achieving good, safe drinking water supplies. Wherever it is adopted, local circumstances will affect the approaches ultimately taken and the pace of implementation.

1.2. The Goal

The Goal of the Bonn Charter is:

> "*Good safe drinking water that has the trust of consumers*".

1.3. Who is the Charter for?

The Charter has been prepared for all those collectively contributing to the provision of safe drinking water from catchment to consumer, recognising the different roles of different parties. Given their responsibility for delivery of safe, reliable drinking water, the Charter recognises the pivotal role of water suppliers for managing those aspects of water supplies within their control.

1.4. How should the Charter be used?

The Charter sets out the principles of an effective drinking water quality management framework and the responsibilities of key parties. It is therefore a basis for designing management and operational systems to ensure a high standard of performance by all. Independent third party assessment is seen as an important aspect of the Charter.

1.5. Impetus for development of the Charter

There has been a growing consensus among water professionals about the need for a more consistent framework to be developed within which the quality of drinking water can be assured. These discussions have been given impetus by the World Health Organisation's development of the 3rd Edition of its *Guidelines for Drinking Water Quality* that place greater emphasis on proactive risk-based management of drinking water supplies, complementing end-product monitoring of compliance against quality standards (Note 2).

Accordingly, a generic framework for the effective management of water quality was developed at an invited workshop of senior water industry professionals in Bonn in October 2001 and refined at a second workshop in February 2004. The Bonn Charter for Safe Drinking Water is the outcome of those discussions. An Implementation Guide – also linked with the WHO Guidelines – will be developed.

2. CONSUMER OUTCOMES

The primary beneficiary of this Charter is the community being served through its water supply system. There are a number of fundamental objectives for which all those involved in the supply of drinking water should strive:

(1) Access to good, safe, and reliable drinking water. This is one of the most basic needs of human society. In many areas water quality may already be high and continuing to improve. In others, where waterborne disease or other quality deficiencies are still prevalent, the basic provision of safe and good supplies is vital.
(2) Water that is not just safe to drink but considered of good aesthetic quality by consumers.
(3) Water supplies in which consumers have confidence.

3. KEY PRINCIPLES

This Charter is based on the identification of key principles considered essential in creating a management framework for the reliable provision of good, safe drinking water. Each of these is considered fundamental; each needs to be addressed.

(1) Management of the whole water supply chain should always be set in the context of management of the whole water cycle, including, but not limited to:
   - Management of water resource provision, including, where necessary, resource augmentation.
   - Management of water and land interactions, taking into account agricultural practices and urban development.
   - The collection and treatment of wastewater.

(2) Systems to ensure drinking water quality should not be based solely on end-of-pipe verification (testing against predetermined standards). Rather, management control systems should be implemented to assess risks at all points throughout water supply systems and to manage such risks.

(3) Such an integrated approach requires close co-operation and partnership between all stakeholders including governments, independent regulatory authorities, water suppliers, local public authorities, health agencies, environmental agencies, land users, contractors, plumbers and manufacturers of relevant materials and products, and consumers themselves.

(4) Open, transparent and honest communication between all stakeholders is essential to developing trust. It contributes to the development of effective water supply systems.

(5) The roles and responsibilities of the different institutions contributing to the delivery of safe and reliable drinking water need to be clearly defined and ensure complete coverage of the system from catchment to consumer. Governments should establish the legal and institutional arrangements necessary to assign appropriate responsibilities among the various parties.

(6) The way in which decisions are made relating to standards for the quality and reliability (Note 3) of water supplies should be transparent.

(7) Water should be safe, reliable and aesthetically acceptable. In progressively realising the goals, however, the standards applied may legitimately vary from location to location and over time.

(8) The price of water should be set so that it does not prevent consumers from obtaining water of sufficient quantity and quality to meet fundamental domestic needs (Note 4).

(9) Any system for assuring drinking water quality should:
   - Be based on the best available scientific evidence.
   - Be sufficiently flexible to take account of the different legal, institutional, cultural and socio-economic situations of different countries.

4. DIAGRAMMATIC REPRESENTATION

This Charter proposes a Framework for the delivery of safe and reliable drinking water, incorporating:

- Development of drinking water safety plans to assess risk throughout water supply systems and mitigate these risks.
- Measurement of drinking water quality against relevant standards.

To be effective, these systems need to exist within an appropriate framework in which roles and responsibilities are clear and where the flow of key information between stakeholders is assured. These relationships are represented below.

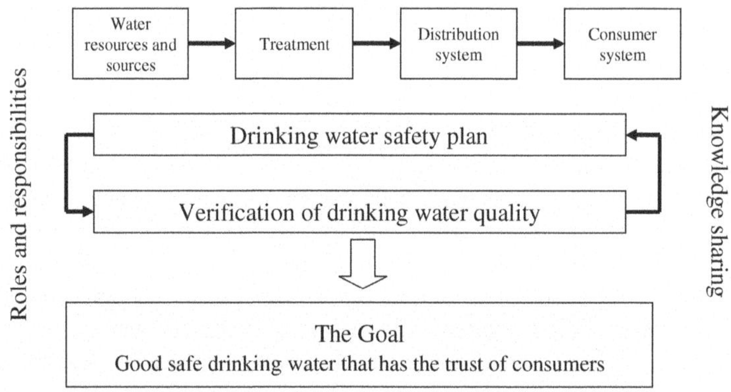

5. ROLES AND RESPONSIBILITIES

An effective water quality management system involves all stakeholders, albeit playing different roles.

5.1. What should Governments do?

Governments are responsible for establishing the legal and institutional framework to ensure delivery of safe and reliable drinking water. To establish such a framework Governments will need to:

(1) Create independent system(s) of co-ordinated water quality regulation that are credible and proportionate.
(2) Take action to protect raw drinking water resources.
(3) Ensure that institutions and coordination arrangements exist to address water quality risks that may occur from the catchment to the tap.
(4) Ensure that these institutions are able to secure sufficient resources to undertake their responsibilities. This would include the creation of mechanisms for funding the installation and ongoing maintenance of water collection, treatment and distribution (Note 5) infrastructure.
(5) Ensure that verification of procedures and water quality outcomes are undertaken with sufficient frequency and that the results are publicly available and transparent.
(6) Establish general accounting policies that ensure water suppliers maintain adequate and auditable accounts.
(7) Establish social policies to ensure all members of society can obtain the services that should be the right of every citizen.

5.2. What should water suppliers do?

Water suppliers (Note 6) play a pivotal role in delivering safe and reliable water to consumers. They should:

(1) In conjunction with partners, develop and implement drinking water safety plans, covering catchment to consumer, and regularly verify their implementation and effectiveness using appropriate operational controls and monitoring.

(2) Put in place systems for testing the quality of water supplied, including those necessary to meet regulatory requirements, and make the compliance results available to the public.
(3) Ensure that the full cost of service provision is identified (including the maintenance and replacement of assets) and that appropriate investments are made in the provision of water services, in line with Government established frameworks for financing.
(4) Ensure staff with sufficient skills and training are available to those involved in the management of each element of the quality process from catchment to consumer.
(5) Maintain adequate and auditable accounts in line with government requirements.

5.3. What should regulatory authorities do?

Independent and credible water regulation is vital to building consumer confidence in the safety and reliability of supplies. (Note that these functions do not need to be the responsibility of just one authority; multiple authorities could be established to fulfil these roles). Regulatory Authorities should:

(1) Establish a regulatory regime that incorporates health and reliability standards for drinking water supplies based on best available scientific evidence and consumer requirements;
(2) Ensure that the process of setting quality- and health-related standards is transparent and the choices made with regard to the level of risk considered appropriate fully disclosed;
(3) Establish verification systems to assess compliance with drinking water quality standards, drinking water safety plans, and to ensure that the results of water quality testing are valid; and
(4) Ensure there is full reporting of compliance results and problems in a way that can be understood by all.

5.4. What should consumers do?

(1) Operate and maintain the water systems in houses in a manner consistent with the maintenance of good water quality. Among other things, this requires the use of appropriate materials (Note 7); and
(2) Be good citizens by ensuring that their behaviour minimises the potential for contamination of water sources to occur or for supply quality or reliability to be diminished.

5.5. What should all parties do?

Share information pertaining to the achievement and maintenance of drinking water quality through open dialogue.

6. DRINKING WATER SAFETY PLANS

A Drinking Water Safety Plan, as described in the WHO Guidelines, is a documented plan (or a number of plans - Note 8) that identifies credible risks from catchment to consumer, prioritises those risks and puts in place controls to mitigate them. It also requires processes to verify the effectiveness of the management control systems put in place and the quality of the water produced. There are 3 key stages that underpin effective Drinking Water Safety Plans:

(1) A system wide assessment of risks from catchment to consumers' taps;
(2) Identification and monitoring of the most effective control points to reduce identified risks; and
(3) Development of effective management control systems and operational plans to deal with both routine and abnormal operating conditions. Recognition must be given to the potential for serious events to occur and provision made for management of these events.

Assessment of the efficacy of the management control systems is also essential. Necessary elements include:

(4) Development of effective measures to assess the effectiveness of the controls that have been put in place and their incorporation within the plan; and
(5) Independent third party verification of the effectiveness of the control systems. The purpose of such verification is to provide confidence that the systems are effective. Irrespective of how such verification is achieved, it should not conflict with the direct management responsibility of water suppliers and others.

Furthermore:

(6) Management control systems should include:
    - A definition of responsibilities;
    - Documented procedures;
    - A training plan to ensure appropriate skills are available among all key operational and other staff.
(7) Management control systems should be appropriate to the size and complexity of the water supply system. For smaller supply systems the use of a more generic package or tool kit may be appropriate.

BOX 1. Water Safety Plans as Described in Who Guidelines

The following is an abstract from WHO Guidelines for Drinking Water Quality (3rd Edition) on development of Drinking Water Safety Plans. It is intended that this Charter and the Guidelines should be in harmony with regard to such plans, and development and verification of water quality standards (see Box 2).

- Health-based targets based on an evaluation of health concerns.
- System assessment to determine whether the drinking-water supply (from source through treatment to the point of consumption) as a whole can deliver water of a quality that meets the health-based targets.
- Operational monitoring of the control measures in the drinking-water supply that are of particular importance in securing drinking-water safety.
- Management plans documenting the system assessment and monitoring plans and describing actions to be taken in normal operation and incident conditions, including upgrade and improvement, documentation and communication.
- A system of independent surveillance that verifies that the above are operating properly.

7. VERIFICATION OF DRINKING WATER QUALITY

In line with the principle of giving greater emphasis to the identification and management of risks from the catchment to the consumers' taps, the number of parameters covered by statutory standards can be kept to a minimum. However, where such standards are necessary they need to be rigorously designed and applied. End product testing should therefore be based on:

(1) A core set of parameters tailored to meet local needs;
(2) Suitable indicator microbiological and chemical parameters where scientifically valid; and
(3) Appropriate monitoring and reporting systems.

The parameters used can be considered in two sets:

(1) Those which give early warning of failure of the management control plan and indicate an immediate risk to health or other serious quality deterioration. These are largely operational control parameters, such as turbidity, and should be accompanied by operational reaction procedures;
(2) Those that are concerned with chronic health or other longer term effects.

Furthermore:

(1) The use of operational control parameters is essential to allow operators to monitor the effectiveness of treatment (e.g. chlorine residual to monitor disinfection, and turbidity to monitor the effectiveness of particle removal); and
(2) Standards should be based on the protection of human health and consumer acceptability.

BOX 2. Setting and Verifying Drinking Water Quality Standards

The WHO Guidelines for Drinking Water Quality (3rd Edition) set out a recommended process for determining drinking water quality standards and verifying performance against them. The Guidelines express the approach required as set out below.

Framework for Safe Drinking Water

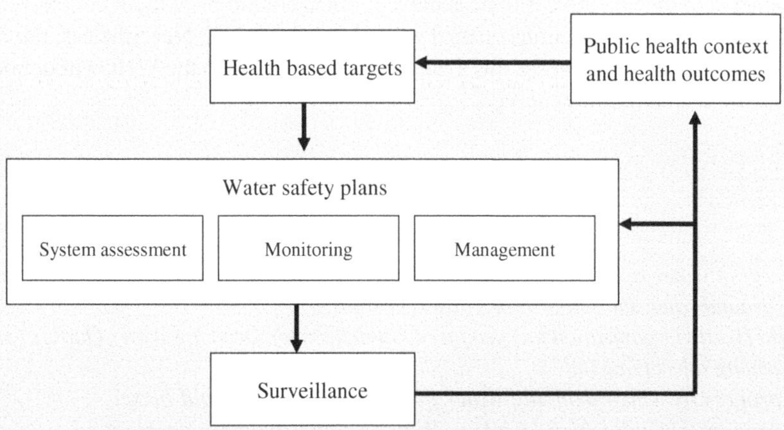

Drinking-water safety is secured by application of a Water Safety Plan (WSP), which includes monitoring the efficiency of control measures using appropriately selected determinands. In addition to this operational monitoring, a final verification of quality is required.

Verification is the use of methods, procedures or tests in addition to those used in operational monitoring to determine if the performance of the drinking-water supply is in compliance with the stated objectives outlined by the Water Quality Targets (WQT) and/or whether the WSP needs modification and revalidation.

For microbial water quality, verification is likely to include microbiological testing. In most cases, it will involve the analysis of faecal indicator micro-organisms, but in some circumstances it may also include assessment of specific pathogen densities. Verification of the microbial quality of drinking-water may be undertaken by the supplier, surveillance agencies or a combination of the two.

Assessment of the adequacy of the chemical quality of drinking-water relies on comparison of the results of water quality analysis with guideline values.

For additives (i.e. chemicals deriving primarily from materials and chemicals used in the production and distribution of drinking-water), emphasis is placed on the direct control of the quality of these products. In controlling drinking-water additives, testing procedures typically assess the contribution of the additive to drinking-water and take account of variations over time in deriving a value that can be compared with the guideline value.

8. CONCLUSION

In combination with the 3rd edition of the *WHO Drinking Water Quality Guidelines* the Bonn Charter provides a comprehensive approach to the elements required to deliver good, safe drinking water that has the trust of consumers.

Access to good, safe drinking water should be the right of every human. While it is acknowledged that economic conditions, civil unrest, drought and other circumstances limit society's capacity to achieve this objective, the adoption of the objective itself is of fundamental importance.

This Charter, however, goes beyond mere adoption; it sets out elements, the implementation of which are considered fundamental if the objective is to be achieved. Local conditions will, of course, remain important and will affect the shape of the institutions created and other steps taken. Nevertheless, the commitment of all parties to implementing the elements of this Charter and responding to the WHO Guidelines will provide important benefits to the communities served.

NOTES

(1) *The concept of water resources in this document refers to raw water collection, whether from surface sources, groundwater, the sea or other brackish sources.*

(2) *The World Health Organisation has produced Guidelines for Drinking Water Quality (3rd Edition) that specify, among other things:*

- *The process by which drinking water quality standards should be set.*
- *The process through which Drinking Water Safety Plans are developed.*

- *The need for independent regulation.*

This Charter complements the WHO Guidelines. Updates in those Guidelines should be taken as updating this Charter in as much as the Charter refers to the Guidelines.

(3) *The term 'reliability' refers to both the quality of the water supplied, and the dependability of supplies.*

(4) *This clause specifically does not suggest how such a price should be set nor should it be taken to mean that those capable of paying the full cost-reflective price should not do so. Rather, it means the most economically disadvantaged should not be excluded from access to water because they cannot afford it. Prices for the economically disadvantaged can be kept at an appropriate level through a range of mechanisms, including transparent cross-subsidies or social subsidies from government. The way by which price is set is a matter for governments, or the regulatory authorities which have delegated authority, as is determination of the groups in society eligible to benefit from such subsidies.*

(5) *In the Bonn Charter the term 'distribution system' should be taken to mean those systems used to deliver water to consumers, whether through pipes, tankers or other means.*

(6) *The term 'Water Suppliers' should be taken to include public and private suppliers, including local public authorities.*

(7) *The system the consumer should be responsible for is that which is used to provide water to a household beyond the point of delivery from the water supplier or vendor.*

(8) *There may be a single plan for a water supply system or multiple, integrated plans where various players have different responsibilities (e.g. there could be one plan for a catchment area, and another for the distribution system). Regardless of the approach taken, it is essential that ownership of the plan(s) is clear and responsibilities allocated appropriately.*

Appendix 2

The Bonn Network

The Bonn Network is a group of water suppliers from across the world who are committed to providing good safe drinking water and who have adopted the principles of the Bonn Charter for Safe Drinking Water. Members of the Network cooperated in the development of best practice through a range of means, including web-based discussions, best practice workshops and newsletters. Network members also supported research carried out by the Network Research Partner, the Centre for Water Science, Cranfield University, UK.

The initial Network has now finished its work and will evolve into an IWA specialist group. Founding Network Members were:

- Three Valleys Water, United Kingdom
- South West Water, United Kingdom.
- Waternet, Netherlands.
- Lyonnaise Des Eaux, France.
- Agbar Group, Spain.
- Águas de Portugal, Portugal.
- Seattle Public Utilities, United States.
- Spanish Town, Jamaica.
- SABESP, Brazil.
- Rand Water, South Africa.
- National Water & Sewerage Corporation, Uganda.
- Ranhill Utilities, Berhad, Malaysia.
- Water Corporation, Australia.
- South Australian Water Corporation, Australia.
- Sydney Water, Australia.

IWA Publishing's authorised EU representative for General Product Safety Regulations is Diane D'Arras, 15 rue Duret, 75116 Paris, France, e-mail: safety@iwap.co.uk.

Printed and bound by CPI Group (UK) Ltd, Croydon, CR0 4YY
12/05/2026
02108863-0009